Springer Series in Reliability Engineering

More information about this series at http://www.springer.com/series/6917

Torben Jespen

ATEX—Explosive Atmospheres

Risk Assessment, Control and Compliance

Springer

Torben Jespen
Koege
Denmark

ISSN 1614-7839 ISSN 2196-999X (electronic)
Springer Series in Reliability Engineering
ISBN 978-3-319-31366-5 ISBN 978-3-319-31367-2 (eBook)
DOI 10.1007/978-3-319-31367-2

Library of Congress Control Number: 2016935607

Printed on acid-free paper

This Springer imprint is published by Springer Nature
The registered company is Springer International Publishing AG Switzerland

Preface

When you cannot avoid the presence of a potentially explosive atmosphere at the plant, or you intend to manufacture and/or place on the market in EU, equipment to be used or installed in explosive atmospheres, you **must** acquire a sufficient knowledge about the **ATEX** subject. This includes both the legislative aspects (especially the two ATEX directives) and all the relevant technical and chemical matters, together with useful guidelines from appropriate standards.

(ATEX stands for EXplosive ATmospheres, shaped from the French translation).

The level of knowledge must at least be sufficient for you to realize when to consult ATEX experts, and to be in a position to communicate with them. In the end, you are the one who will be responsible for legal compliance, as far as the authorities are concerned.

One thing is clear: Familiarizing oneself with the EU directives (concerning the workplace and the equipment) is not enough because the legislation does not tell you exactly what to do. It provides you primarily with important guiding principles, for example about what to do when the possibility of an explosive atmosphere is high, middle, or low, but the legislation does not fix numbers to these levels of likelihood.

Nobody can guide you in detail on all the practical aspects of ATEX, but if you have acquired a certain understanding of the whole subject, the principles and the important tools, standards and technical aspects, you will be in a good position to fulfill your legal obligations in a qualified manner. In many cases you can do it on your own—but at least when being supported by the relevant experts.

The employer being responsible for compliance with the ATEX Workplace directive 1999/92 for the safety and health of his employees needs to have some knowledge about ATEX equipment which is regulated by the ATEX equipment directive 2014/34. And the manufacturer of ATEX equipment, must most certainly also be familiar with the content of the Workplace directive, and be able to guide his customers appropriately.

This book aims at fulfilling the task of providing both the employer, who has to tackle ATEX scenarios at his plant, and the manufacturer/importer of ATEX equipment, with basic ATEX knowledge that will allow them to solve simple ATEX problems by themselves, and will guide them towards appropriate solutions where more difficult problems have to be managed.

Finally, the author wishes to express his gratitude for many helpful and inspiriting discussions and comments received from colleagues at work, especially Mr. Sigfred K. Nielsen (Bachelor of Science in Mechanical Engineering) who was always ready to study and comment on any draft material prepared by me.

Koege, Denmark
January 2016

Torben Jespen

Contents

Chapter 1
ATEX—Introduction

Safety (i.e. absence of unacceptable risks) in areas where potentially explosive atmospheres may occur, is beyond doubt one of the most difficult subject to tackle in the field of occupational health and safety.

Not only are the involved technical and chemical aspects generally for experts to manage, especially when you have to delve deeply into these matters. But also the EU ATEX (Explosive Atmosphere) legislation that you have to comply with, may leave you with a lot of unanswered questions because the requirements are based on principles, and although you seek supporting guidance, as for instance through the specifications in relevant standards, you may still be in doubt.

Are you a manufacturer of equipment intended to be installed in areas where explosive atmospheres may be present, or are you an employer whose activities include such areas, you must possess (yourself or some of your employees) the basic knowledge about the ATEX legislation and related technical matters. This is necessary in order to meet the legislative obligations—or at least to understand when it is necessary and appropriate to consult external ATEX experts, and be able to communicate with and understand the inputs from those experts.

The present guide aims at fulfilling this objective.

In short, you could say that the legislative safety approach (where the occurrence of explosive atmospheres cannot be avoided, e.g. by inerting) is based on the principle, that the higher the likelihood for an explosive atmosphere being present, the more stringent the requirements on the equipment (to be installed in these areas). Stringent requirements mean less likely to be able to ignite an explosive atmosphere (i.e. the less likely its own possible ignition sources will trigger off).

Therefore, the employer must assess the likelihood for an explosive atmosphere being present on his premises, which according to ATEX workplace directive 1999/92 must be done utilizing the zone classification system which defines and ranks such hazardous areas in zones using "elastic" wording as: Continuously present, likely to occur and not likely to occur.

It is worth emphasizing that this zone classification scheme is only about the likelihood of an explosive atmosphere being present **without** taking into

T. Jespen, *ATEX—Explosive Atmospheres*, Springer Series
in Reliability Engineering, DOI 10.1007/978-3-319-31367-2_1

consideration the risk regarding the possibility of ignition sources being present and the likely outcome, should an explosion become a reality (aspects, which put together constitutes the explosion risk assessment).

Having accomplished a sound zone classification, the employer must then select equipment according to the criteria laid down in ATEX workplace directive 1999/92, unless the specific explosion risk assessment indicates otherwise. Again, the legislation leaves it very much to the employer's own judgment to arrive at the appropriate decision concerning safety measures.

In general, the legislation builds very much on likelihood considerations, and therefore, although you meet the legal obligations, an explosion may happen. In many cases the possibility of an explosion has to be accepted for example because the incoming process flow may contain un-removable ignition sources, which means that supplementary mitigating measures must be implemented in order to protect the affected employees (and the equipment).

Part I on workplace safety, consisting of the Chaps. 2–11, sets focus on the abovementioned aspects.

The zone classification task is the essential (and often very difficult) foundation for all the safety precautions imposed on the employer at the prevention and protection against the risk due to the possible occurrence of explosive atmospheres. In order to create a reasonable good knowledge about these aspects, the zone classification concerning the explosive fuels: Oil fluids, natural gas and dusts (especially coal dust), are dealt with in the Chaps. 8–10, respectively.

Manufacturers of products intended for use in possible explosive atmospheres must comply with the ATEX equipment directive 2014/34, which groups and categorizes equipment according to the intended use and the level of protection against own possible ignition sources becoming effective. This clearly facilitates the customer's choice of appropriate equipment according to the legal requirements placed on him by ATEX workplace directive 1999/92.

In order for manufacturers to be able to guide customers about equipment selection and other important and difficult ATEX questions for the user, the successful manufacturer must possess a sufficient knowledge about the aspects covered by the ATEX workplace directive 1999/92 and related guidelines. So, although Part II on equipment safety, consisting of the Chaps. 12–16 and dealing with the ATEX equipment directive 2014/34 and related documents, is of special importance for the manufacturer of equipment, he should most certainly be familiar with the main aspects covered by Part I (Chaps. 2–11).

Mechanical equipment (which is generally without effective ignition sources) intended for use in ATEX classified zones require special attention in order to demonstrate this fact, or alternatively to determine the likelihood of a possible own ignition source to manifest itself.

In parallel with the employers' zone classification, the manufacturer must perform a probability assessment: Is the potential ignition source prevented during normal operation, does the prevention also include situations with expected malfunctions or are even rare malfunctions included in the prevention?

This probability assessment, leading to the categorization of the equipment takes place under the heading: Ignition Hazard Assessment (IHA), which is the subject of Chap. 13.

Important ATEX equipment directive 2014/34 aspects, as the Essential Health and Safety Requirements (EHSR) required for directive compliance according to equipment category, conformity assessment procedures and markings of the equipment are dealt with in the Chaps. 12 and 16.

Finally, Part III (Chap. 17) contains some basic knowledge on fires and explosions, which should be of interest for those who inquire for a little deeper insight regarding the technical and chemical aspects behind the explosion parameters.

1.1 Disclaimer

The purpose of this guide is to provide general guidance on issues arising out of the EU ATEX legislation—especially on zone classification, explosion risk assessment, product categorization, Ex-marking and related technical/chemical aspects.

The author has endeavored to ensure the accuracy of the guidelines. But legislative issues are subject to change, and the interpretation may differ over time and between experts, and this guide cannot be exhaustive in its coverage of all the aspects. The same goes for some of the more technical and scientific subjects that are closely connected to the ATEX legislation.

Anyhow, the guide intends to reflect the interpretations and recommendations regarded as valid at the time it was published. But supplementary information may be needed depending on the circumstances, and it must therefore be emphasized that this text is intended to guide you in the right direction, but cannot be relied upon as a guarantee for being in exact compliance with the EU legislation or for not failing when a specific decision has to be taken.

In the end, being in compliance with EU legislation concerning a specific product is first of all the responsibility of the product manufacturer and implies the accomplishment of a sound specific explosion risk assessment for which he may need to be supported by professional advice procured from outside his company.

The same holds for the employer when hazardous ATEX areas (zone classified) are present at his premises.

Part I
ATEX Workplace Safety Aspects—Employer Obligations

Chapter 2
ATEX—Workplace Legislation

2.1 Introduction

ATEX (shaped from the French "ATmosphères EXplosibles") is the name commonly given to the two EU directives aiming to control the risks caused by explosive atmospheres: Directive 1999/92 (ATEX Workplace directive) and Directive 2014/34 (ATEX Equipment directive).

Directive 1999/92 is a directive laying down minimum requirements (employer obligations) for protecting health and safety at work, especially through zone classification of hazardous areas, i.e. areas where explosive atmospheres may occur, and through explosion risk assessments and appropriate ATEX equipment selection.

Directive 2014/34 is a Product directive harmonising the Essential Health and Safety Requirements (EHSR) to be complied with by the manufacturers of ATEX equipment, including instructions for equipment categorisation, conformity assessment procedures and CE- and Ex-marking obligations.

An "**explosive atmosphere**" in the scope of the two directives is defined as "a mixture with air, under atmospheric conditions, of flammable substances in the form of gases, vapours, mists or dust in which, after ignition has occurred, combustion spreads to the entire unburned mixture".

Atmospheric conditions (be it inside or outside process equipment) are generally regarded (but not defined in the directives) as places having a pressure between 0.8 bara and 1.1 bara, and a temperature between −20 and +60 °C. Areas outside "atmospheric conditions" do not fall under the scope of the two directives.

Although the operating conditions at a plant in general are outside "atmospheric conditions" and therefore fall outside the scope of the directives, start-up and shut-down situations may very well belong to "atmosphere conditions" and must therefore comply with the requirements of the directives.

Of course, an employer cannot ignore the existence of a potential explosive atmosphere under non-atmospheric conditions when fulfilling his obligations regarding health and safety at work, e.g. under the scope of the Framework

T. Jespen, *ATEX—Explosive Atmospheres*, Springer Series
in Reliability Engineering, DOI 10.1007/978-3-319-31367-2_2

directive 89/391 and the Chemical agents directive 98/24, but exact ATEX directive compliance is not required.

Besides the ATEX Equipment directive 2014/34, compliance with other product directives, as for instance the Machinery directive 2006/42 and the Electromagnetic Compatibility directive 2014/30, may be relevant for the manufacturer.

Note: Products for use in potentially explosive atmospheres are explicitly excluded from the scope of the Low Voltage directive 2014/35.

Note: Equipment intended for use in domestic and non-commercial environment where explosive atmospheres may only rarely be created, solely as a result of accidental leakage of fuel gas, are excluded from the scope of the ATEX Equipment directive 2014/34.

ATEX Equipment directive 2014/34 is supported by several harmonised standards offering the manufacturer the possibility to obtain the presumption of conformity with the requirements of the directive, see subsequent Chap. 5.

In general, workplace directives are not directly supported by EN standards, but the following three standards may be useful when you endeavour to comply with the ATEX Workplace directive 1999/92:

- EN 1127-1: Explosive atmospheres—Explosion prevention and protection—Basic concepts and methodology
- EN 60079-10-1: Explosive atmospheres—Classification of areas—Explosive gas atmospheres
- EN 60079-10-2: Explosive atmospheres—Classification of areas—Combustible dust atmospheres

The ATEX Workplace directive 1999/92 and the above-mentioned area classification standards do not contain very detailed practical guidance on the zone classification of hazardous areas, and without further studies, practical experience and/or expert guidance, the zone classification task may seem as an attempt to comply with legislation of a "rubber-like" nature where individual viewpoints could differ considerably.

For some important areas further guidance can be obtained from "codes of good practice" as the following, focusing on flammable fluids and natural gas, respectively:

- IP 15: Area classification code for installations handling flammable fluids, published by the Energy Institute, London, 2005 and
- IGEM/SR/25: Hazardous area classification of natural gas installations, Edition 2, Communication 1748, Published by The Institution of Gas Engineers and Managers, Kegworth, 2010.

In order to acquire a reasonable knowledge and understanding of the important aspects falling within the ATEX sphere, it is recommended first to acquaint oneself with the basic obligations under the scope of the ATEX Workplace directive 1999/92. Although you may be primarily a manufacturer/supplier of ATEX equipment, it is essential that you understand your customers' situation, are able to provide some "expert" guidance to them, and know the importance of having agreed on the appropriate (categorized) equipment to be supplied when signing the final contract.

Especially because of the complexity of the ATEX field, the manufacturer often—more or less explicitly—acts as an adviser/consultant who indirectly may become responsible in relation to the ATEX Workplace directive.

On the other hand, the user (employer) may easily become a manufacturer if he combines ATEX equipment into a final assembly of his own, because the ATEX Equipment directive (in line with the Machinery directive) also covers putting into service.

2.2 Explosions—Hazards and Risks

Chapter 17 contains a more general introduction to the "Fire and explosion risks" subject, so only a few words should suffice here in order to introduce the hazards and risks which the ATEX directives aim at controlling.

A **fire hazard** requires the simultaneous presence of three elements (removing one of these elements eliminates the hazard), often called the "**fire triangle**":

- Fuel to feed the fire,
- Air (oxygen), and
- An effective ignition source.

A **fire risk assessment** covers three aspects: The likelihood of the fire hazard to be present, the likelihood of the fire hazard triggering and the likely outcome of this event.

In order to create **an explosion hazard** two more requirements must be fulfilled at the same time (extending the "fire triangle" to an "**explosion pentagon**"):

- A premixing of the fuel and air (oxygen) in the right proportions, i.e. the fuel concentration must be between the Lower (LEL) and the Upper Explosive Limit (UEL)
- A sufficient confinement of this mixture for the sudden release of energy to result in temperature increases and concomitant significant pressure increases

Without confinement, heat transfer could not occur rapidly enough to allow continued propagation of the "explosion" which instead will become something like a large fireball with no appreciable forces associated with it. Such unconfined explosions, also called flash fires, can be equally destructive as an explosion due to the secondary fire risk and the potential for disturbance of powder layers leading to secondary explosions.

An **explosion risk assessment**—in line with a fire risk assessment—covers three aspects: The likelihood of the explosion hazard to be present, the likelihood of the explosion hazard triggering and the likely outcome of this event.

2.3 Workplace Obligations (Directive 1999/92)

Note: Mineral extracting industries covered by Directive 92/91 or Directive 92/104 do not fall under ATEX Workplace directive 1999/92.

Note: The following overview of obligations placed on the employer by the ATEX Workplace directive 1999/92 does not cover all the requirements in details.

First of all, the employer shall adhere to the basic principles for explosion prevention and protection, in order of priority:

- Prevent the formation of explosive atmospheres (see Chap. 4), or where this is not a reasonable possibility,
- Avoid the ignition of the explosive atmospheres, and
- Mitigate the detrimental effects of an explosion so as to ensure workers health and safety.
- Where appropriate, these measures to be combined and/or supplemented with measures against the propagation of explosions.

Abovementioned first three principles also apply to ATEX equipment—where it is referred to as the principles of **integrated explosion safety**.

In order to comply with these principles, the employer must embark on the explosion risk assessment, which is initiated by determining whether a hazardous explosive atmosphere can arise under the circumstances in question, including normal operation and maintenance, commissioning and decommissioning, foreseeable malfunctions and failures and foreseeable misuse.

Where the possible presence of an explosive atmosphere cannot be excluded, the employer must perform (or have elaborated based on expert consultations) a full **explosion risk assessment** covering:

- The likelihood that explosive atmospheres will occur and their persistence,
- The likelihood that ignition sources will be present, become active and effective (i.e. able to ignite the explosive atmosphere in question),
- The possible interactions with the surroundings, and
- The scale of the anticipated effects.

The explosion risk assessment must be documented in the obligatory **Explosion protection document**.

Where the hazard identification and risk assessment of an area (place) points to the fact that an explosive atmosphere is not expected to occur in such quantities as to require special safety precautions, the area is regarded as a non-hazardous area concerning ATEX aspects, and no further action is required in that connection.

Where the hazard identification and risk assessment of an area (place) points to the fact that an explosive atmosphere may occur in such quantities as to require special safety precautions, the area is regarded as a hazardous area concerning ATEX aspects (and subject to zone classification).

Generally speaking, unless the volume of the potential explosive atmosphere is very small and/or its persistence of a very short duration, the area should be regarded as a hazardous area.

The subsequent parts contain more information and comments concerning the abovementioned aspects.

2.3.1 Zone Classification—Likelihood of Explosive Atmosphere Being Present

The first step of the obligatory workplace explosion risk assessment (to be documented through the Explosion protection document) consists of an assessment of the likelihood that an explosive atmosphere will occur and its persistence.

Areas, where its occurrence cannot be neglected and special precautions are required, shall be classified in terms of zones based on the frequency and duration of the occurrence of the explosive atmosphere. The zone classification is therefore a way of expressing the likelihood of an explosive atmosphere being present in an area.

Note: Preventing the formation of explosive atmospheres, if possible, must of course—in compliance with the legislation—be the first step to be taken, see Chap. 4.

ATEX Workplace directive 1999/92 defines the zones as stated in Tables 2.1 and 2.2, distinguishing between flammable gas, vapor and mist zones and dust zones.

Note: Area classification does not apply to catastrophic failures (e.g. rupture of a storage tank or a pneumatic conveyor), which are beyond the concept of abnormality dealt with in the assessment. However, normal operation includes all conditions within the design parameters of the plant, including start-up, stop and failure of fragile components.

The zone definitions communicate to you that the possibility (risk) of an explosion hazard being present increases through the zones 2 to 1 to 0 and through the zones 22 to 21 to 20, signaling a need for increasing precautionary matters. But

Table 2.1 Flammable gas, vapor and mist

Zone	Definition
Zone 0	A place in which an explosive atmosphere consisting of a mixture with air of flammable substances in the form of gas, vapor or mist is present continuously or for long periods or frequently
Zone 1	A place in which an explosive atmosphere consisting of a mixture with air of flammable substances in the form of gas, vapor or mist is likely to occur in normal operation occasionally
Zone 2	A place in which an explosive atmosphere consisting of a mixture with air of flammable substances in the form of gas, vapor or mist is not likely to occur in normal operation but, if it does occur, will persist for a short period only

Table 2.2 Combustible dust

Zone	Definition
Zone 20	A place in which an explosive atmosphere in the form of a cloud of combustible dust in air is present continuously, or for long periods or frequently
Zone 21	A place in which an explosive atmosphere in the form of a cloud of combustible dust in air is likely to occur in normal operation occasionally
Zone 22	A place in which an explosive atmosphere in the form of a cloud of combustible dust in air is not likely to occur in normal operation but, if it does occur, will persist for a short period only

attempting a zone classification based directly on the definitions could easily lead to endless reflections because the directive does not specify what is meant by "continuously", "long period", "short period", "frequently" and "occasionally".

However, help is available as it is commonly accepted to use the guidance concerning duration and frequency at the zone classification as stated in Table 2.3.

Zone classification takes into consideration the type of explosive mixture and the likelihood of its occurrence (determined by the frequency and duration of its occurrence, which outside equipment is influenced by the degree of ventilation of the area). On drawings it is recommended to indicate the zoning as ilustrated in Fig. 2.1.

Concerning releases of flammable gas, vapor, mist or dust into the atmosphere from a release point (for example through vents and failure leaks), one uses as a "starting point" the concept "Grade of release" which is closely related to the zone classification conception. But the Grade of release is dependent solely on the frequency and duration of the release, and independent of the rate and quantity of the release, the degree of ventilation or the characteristics of the fluid. The Grade of release is classified/defined as stated in Table 2.4.

The terms "Grade of release" and "Zone" are (as underlined above) not synonymous, but in most cases under unrestricted "open air" conditions there exists a direct relationship between "Grade of release" and "Zone" according to:

- A continuous grade normally leads to a Zone 0 or 20
- A primary grade normally leads to a Zone 1 or 21
- A Secondary grade normally leads to a Zone 2 or 22

Table 2.3 Guidance on zone classification

Zone	Duration/frequency guidance
Zone 0 and 20	T > 1000 h/year **or** T > 1 h/shift
Zone 1 and 21	10 h/year < T < 1000 h/year
Zone 2 and 22	1 h/year < T < 10 h/year **and** T < 1 min/shift

T Estimated time period with explosive atmosphere being present

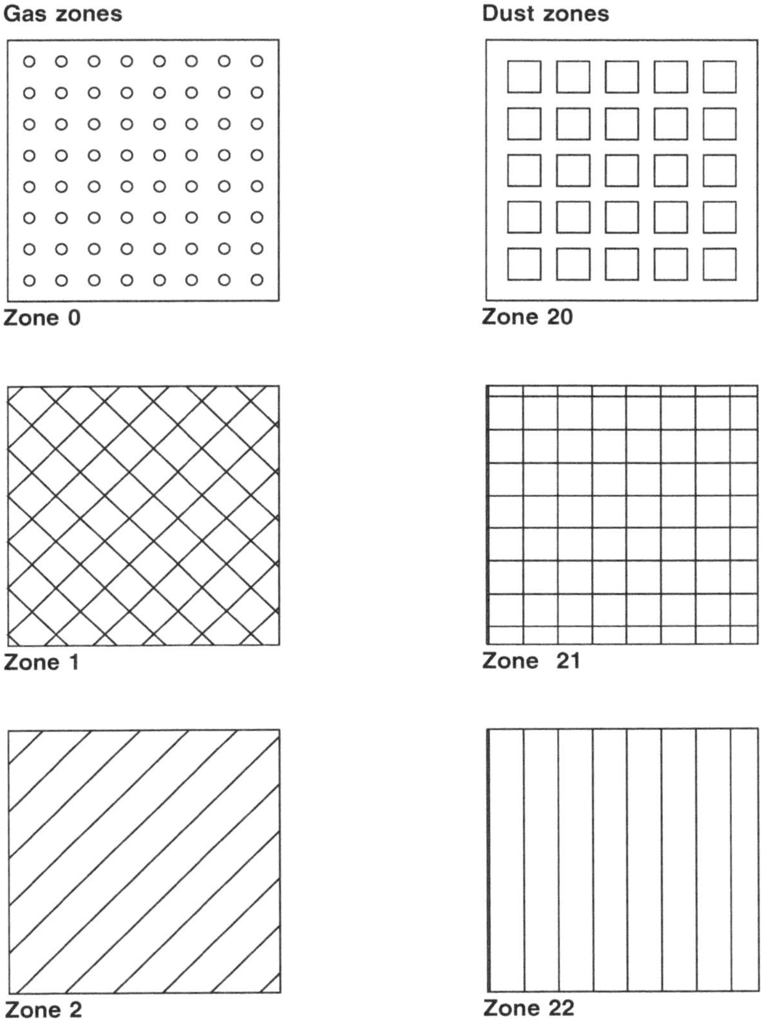

Fig. 2.1 Recommended zone indications on drawings

The general procedure for zone classification can be divided into five steps:

1. Obtain process information, such as temperature, pressure, concentrations, and about plant design, and material explosion risk parameters, such as flash point, limits of explosion and layer ignition temperature—see also Chap. 3.
2. Determine internal zone classifications as appropriate (here the zone extents will be fixed by the equipment).

Table 2.4 The grade of release classes and definitions

Grade of release	Definition	Time guidance (hours/year)
Continuous grade	A release that is continuous or is expected to occur frequently or for long periods	T > 1000
Primary grade	A release that can be expected to occur periodically or occasionally during normal operation	10 < T < 1000
Secondary grade	A release that is not expected to occur in normal operation and, if it does occur, is likely to do so only infrequently and for short periods	1 < T < 10

3. Identify release points and assign grade of releases.
4. Consider relevant effects of ventilation.
5. Determine zone numbers and zone extents.

The difficulties in zone classification are very much dependent on the availability of guidelines, applicable examples, release dispersion model calculations, experiments and/or own practical experience.

Chapters 8–10 set focus on some important aspects and principles attached to zone classification associated with the materials **oil fluids**, **natural gas** and **dusts**, including coal dust, respectively.

If the plant owner (customer/employer) has decided on the final zone classification of the hazardous areas at his plant, a simple equipment supply contract can be based on this classification.

In any case, a zone classification must be available where explosive atmospheres may be present in order to minimize explosion risks based on the correct equipment categories being used or installed in these hazardous areas (and to ensure compliance with the legislation). Because of the difficulties associated to the zone classification task, an equipment or plant department provider should be prepared to present drafts concerning abovementioned documents, which can form the basis for deliberations with a customer in order to tailor a final agreed zone classification, which can ensure the delivery of the appropriate equipment.

In some cases it may not be possible to supply equipment possessing the legally required (sufficiently) low level of own ignition sources, and the possibility of an explosion has to be accepted. Mitigating measures must then be decided upon—see subsequent Sect. 2.5.

In some cases equipment own effective ignition sources can be excluded so that the potential explosive atmosphere does not lead to an explosion hazard due to the equipment. But an effective ignition source (e.g. a stone or piece of metal) may for example enter a hazardous area as an element of the process stream, which flows to the process equipment in question. This is a situation which—most certainly—has to be dealt with in the explosion risk assessment.

If a hazardous area is located inside machinery, the machinery must comply with EHSR (Essential Health and Safety Requirement) 1.5.7 on explosion risks in Annex I to the Machinery directive. Equipment which is put inside this area of the machinery must comply with ATEX Equipment directive 2014/34, but the machinery as such does not fall under the ATEX directive (unless the machinery is intended to be placed in an area where it will be surrounded by or have interface to an explosive atmosphere).

To which extent the manufacturer of machinery has to take into account the possible presence of ignition sources in the process stream, very much depends on the "intended use" of the equipment and foreseeable misuses, and the knowledge a customer reasonably can expect from the supplier—taking the specific circumstances into consideration. In any case, if the supplier has some knowledge about the possible presence of ignition sources in the process stream in question, he is obliged (at least) to warn his customer.

Note: Area classification does not take into account the consequences of ignition.

Note: Zone classified areas must be marked with EX signs at point of entry—where necessary (e.g. at relevant entrance to rooms or areas in which an explosion hazard may arise. If the hazardous area only makes up a part of the whole area, that part may be marked by yellow/black diagonal stripes on the floor).

2.3.2 Likelihood of Ignition Sources Being Present and Effective

As stated in the basic principles for explosion prevention and protection, the possible occurrence of an explosive atmosphere which cannot be eliminated, must lead (2nd priority) to the avoidance of possible ignition sources, which will eliminate the explosion hazard.

Equipment and protective systems for use in zone classified areas must comply with the ATEX Equipment directive 2014/34, which groups and categorizes the equipment according to intended use and the likelihood of own possible ignition sources becoming effective. In general, this equipment categorization, indicating very high, high or normal level of protection can be used directly in equipment selection—taken into account the suitability for gas/vapor or dust. See subsequent Sect. 2.4 and Chap. 12.

Where the explosion risk assessment does not indicate otherwise, and the equipment selection is in conformity with the equipment selection rules of Annex II, Sect. B of ATEX Workplace directive 1999/92 (see subsequent Sect. 2.4), the explosion risk can be regarded as sufficiently minimized, and no supplementary mitigation measures are required.

A more specific assessment of possible ignition sources entering the process stream and may become effective, can be a very complicated task. Therefore, if the possibility of such ignition sources entering the system cannot be disregarded, the conclusion should be, that the possibility of the atmosphere becoming explosive must be eliminated for instance by inerting, and/or mitigating measures must be implemented.

Note: Chapter 14 gives an overview of all the possible ignition sources, which are also dealt with in EN 1127-1 "Explosive atmospheres—Explosion prevention and protection—Basic concepts and methodology".

2.3.3 Anticipated Effects

A risk assessment must always include an estimation of the likelihood of the occurrence (indicated here through zoning and possible presence of effective ignition sources) and the likely/foreseeable harmful effects, should the explosion hazard become a reality, taking any implemented mitigating measure into account.

The anticipated outcome of an explosion will depend on several factors, such as the actual process conditions, plant layout, material explosive properties, expected appearance of effective ignition sources, any mitigating measures and how many people could be harmed.

Possible secondary effects such as the creation of explosive dust atmospheres from layers of dust and the initiation of fires at the plant should also be part of the assessment.

Not only health and safety aspects is generally expected to be included in the assessment of risks, as explosion hazards are normally associated with considerable possible economic consequences.

The overall risk assessment may indicate appropriate deviations from the "standard" scheme for equipment selection (see subsequent Sect. 2.4), put forward in Annex II, Sect. B of ATEX Workplace directive 1999/92. I.e. in some situations equipment on a higher protection level (e.g. "category 2" equipment instead of "category 3" equipment) should be selected than according to the "standard" scheme, and in other situations it is justified to use equipment on a lower protection level than according to the "standard" scheme.

2.3.4 Explosion Protection Document

In order to document legislative compliance and to provide a survey on the control of the ATEX aspects at the workplace, the employer must ensure that an "Explosion protection document" is drawn up and kept up to date, at least demonstrating:

- That an explosion risk assessment has been accomplished,
- That adequate measures will be implemented, fulfilling the aims of the workplace directive,
- The areas which have been zone classified, e.g. on appropriate diagrams or drawings,

- The areas where the minimum requirements set out in Annex II of the workplace directive apply (zone classified areas as appropriate and equipment in non-hazardous areas of significance for equipment located in zones),
- That the workplace and work equipment are designed, operated and maintained with due regard for safety,
- That arrangements have been made for the safe use of work equipment in accordance with Workplace equipment directive 89/655.

The Explosion protection document must be drawn up prior to the commencement of work.

Note: Before a workplace containing zone classified areas is used for the first time, its overall explosion safety must be verified. The **verification** must be carried out by persons competent in the field of explosion protection as a result of their experience and/or professional training. And as a logical conclusion, the verification must be seen as constituting a report to be added to the Explosion protection document.

2.4 Equipment Selection

The likelihood of an explosive atmosphere to occur, expressed through the zone classification, has a decisive influence on the selection of equipment to be used or installed in these areas according to Annex II, Sect. B of ATEX Workplace directive 1999/92.

Logically, the higher the probability of an explosive atmosphere to occur, the higher the needed protection against own effective ignition sources. For group II equipment (equipment intended for use outside underground parts of mines and those parts of surface installations of such mines), the needed protection falls in one of three categories (according to ATEX Equipment directive 2014/34—see also Chap. 12):

- Category 1, possessing a very high level of protection, which are allowed to be used in zone 0 and zone 20 (and all the other zones) provided they are suitable for gas and/or dust, as appropriate.
- Category 2, possessing a high level of protection, which are allowed to be used in zone 1 and zone 21 (and zone 2 and zone 22) provided they are suitable for gas and/or dust, as appropriate.
- Category 3, possessing a normal level of protection, which are allowed to be used in zone 2 and zone 22 provided they are suitable for gas and/or dust, as appropriate.

Equipment and protective systems must be chosen in accordance with the abovementioned categories, **unless** the Explosion protection document based on the risk assessment indicates otherwise.

Note: Abovementioned relationship between zones and categories does not take into account the potential consequences of an explosion, nor supplementary aspects as the toxicity of the materials involved.

Further criteria such as temperature class, type of protection and explosion group must be considered to ensure safe operation of equipment in hazardous areas. These criteria depend on the combustion and explosion properties of the substances used—see also Chaps. 5 and 12.

The necessary and sufficient information for the correct selection of equipment should appear from the marking of the equipment which should be repeated in the ATEX EU Declaration of conformity.

Note: Mechanical equipment without own ignition sources are in principle not covered by the ATEX Equipment directive 2014/34. Therefore, such equipment cannot be accompanied by an ATEX EU Declaration of conformity, but may of course be used or installed in zone classified areas.

Protective systems (separately placed on the market for use as autonomous systems) are covered by ATEX Equipment directive 2014/34, but they are not grouped and categorized as the equipment, and do not have a specific relationship with the zoning system. The essential requirements for protective systems do not depend on the likelihood of the occurrence of an explosive atmosphere or the likelihood of effective ignition sources being present. Protective systems must be designed and dimensioned in such a way as to reduce the effects of an explosion to a sufficient level of safety—regardless of the probability of an explosion.

Protective systems (as well as equipment) may be designed for a particular explosive atmosphere. In this case, they must be marked accordingly.

2.5 Explosion Protection Measures—Mitigating the Effects

Where the possibility of an explosion has to be accepted due to the possible occurrence of equipment own ignition sources or ignition sources in the process flow (which cannot be avoided), the elimination or minimization of the safety risks has to be carefully planned and implemented.

The employer is of course responsible for the safety and health of the employees and must decide on the technical and/or organisational mitigating measures to limit the effects of an explosion to a safe level. But the design and dimensioning of specific equipment is entrusted equipment manufacturers, who must take care of the compliance with relevant product directives and standard specifications.

The technical mitigating measures can be grouped as (see EN 1127-1 for further information):

- Explosion relief
- Explosion resistant design (Explosion pressure resistant design **or** Explosion pressure shock resistant design)

- Explosion suppression
- Prevention of flame and explosion propagation

These mitigating measures are dealt with in more detail in Chap. 6.

Preventive measures as “good housekeeping”, “ventilation” and “inerting” are touched upon in Chap. 4.

Chapter 3
ATEX—Risk Parameters

3.1 Introduction

In order to be able to carry out an explosion risk assessment including an appropriate area zone classification, full details of the plant and its operating conditions are needed.

In addition to this it is necessary to know the important flammability (or explosibility) properties of the materials being handled.

Materials possessing the possibility of creating an explosive atmosphere can be divided into the following groups:

- **Flammable gases** are often compounds of carbon and hydrogen, but hydrogen, and for instance ammonia, also belong to this group. Flammable gases and vapours require—in general—only small amounts of energy to initiate a reaction with atmospheric oxygen.
- **Flammable liquids** (vapours) are often hydrocarbon compounds. Even at room temperature sufficient quantities of these near their surface can often change into vapour phase to form an explosive atmosphere near their surface. Other liquids require increased temperatures for doing so.
- **Flammable mist** consisting of very small droplets with a large overall surface area can be created when spraying a flammable liquid (In such a case, the Flash point (T_F) of the liquid is of lesser importance).
- **Flammable dusts** are particles with diameters less than 0.5 mm, generally of organic materials, but many metals and some non-metallic inorganic materials can burn or explode—if finely divided and dispersed in air in sufficient concentrations.

In general, explosibility aspects are dealt with separately for dust and gas/vapor, with flammable mists as a special case of the explosibility of gases and vapours.

T. Jespen, *ATEX—Explosive Atmospheres*, Springer Series
in Reliability Engineering, DOI 10.1007/978-3-319-31367-2_3

As an introduction it is worth noting the two major differences that exist between gas/vapor and dust explosibility questions concerning the spread of flammable atmospheres and the ignition characteristics.

When gas/vapor is released, the resultant flammable cloud will gradually disperse into the atmosphere. When the gas/vapor concentration has fallen below the lower explosive limit (LEL) there is then no ignition risk. This helps to define zone boundaries with some precision.

In contrast, the release of dust will form a cloud initially, which will settle on to the floor, ledges, etc., from where it could be stirred up again to form a flammable cloud. This means that a potential risk remains, and this has to be taken into account in area classifications. 1 mm layer of dust having a density of 500 kg/m^3 may create a dust cloud of 100 g/m^3 to a height of 5 m, which could constitute an explosion hazard.

Furthermore, it is impracticable to calculate the extents of hazardous zones outside equipment for dust—as employed for gasses and vapours.

Flammable gas/vapor clouds are sensitive to ignition by low energy sources, but their auto-ignition temperatures are generally in excess of 250 °C, with the majority being above 350 °C. It is very unlikely, therefore, that the outer surface of electrical apparatus will get sufficiently hot to cause direct ignition of a flammable gas/vapor/mist cloud. Surface temperatures must of course, be taken into account when selecting apparatus.

Many powders have dust cloud ignition temperatures that are significantly lower than the auto-ignition temperatures of gases/vapours. In addition to this, layers of powder on hot surfaces will often decompose when exposed to temperatures a low as 150–200 °C. This decomposition can lead to smouldering or combustion with flame, producing a potential source of ignition for any flammable dust cloud in the vicinity. This is another aspect that has to be considered in the selection of apparatus.

3.2 Gas/Vapour Explosion Risk Parameters

The key property of a flammable liquid fluid concerning fire and explosion risks is its volatility, expressed through its Flash point (T_F), since it will determine the extent of rapid vapour formation from any release.

3.2.1 Flash Point (T_F)

The Flash point (T_F) is the lowest temperature at which a source of ignition causes the vapour to ignite and the flame propagate across the surface of the liquid.

The concept of T_F applies to all liquids including liquefied gases. In practice, because the temperatures are so low, flash point is not relevant to the permanent

Table 3.1 Oil fluid classes

Class	Fluids
Class 0	LPG (Liquefied Petroleum Gases)
Class I	Having $T_F < 21$ °C
Class II (1)	Having 21 °C ≤ T_F ≤ 55 °C, handled below T_F
Class II (2)	Having 21 °C ≤ T_F ≤ 55 °C, handled at or above T_F
Class III (1)	Having 55 °C < T_F ≤ 100 °C, handled below T_F
Class III (2)	Having 55 °C < T_F ≤ 100 °C, handled at or above T_F
Unclassified	Having $T_F > 100$ °C

T_F Flash point temperature

Table 3.2 Oil fluid categories

Category	Description
A	A flammable liquid that, on release, would vaporize rapidly and substantially
B	Any flammable liquid, not belonging to category A, but handled at a temperature sufficient for boiling to occur on release
C	Any flammable liquid, not belonging to category A or B, but which can, on release, be at a temperature above its flash point, or **form a flammable mist or spray**
G(i)	A typical methane-rich natural gas
G(ii)	Refinery hydrogen

gases such as methane and hydrogen, which are considered to be capable of forming flammable mixtures at any temperature.

T_F can be regarded as a lower temperature limit of flammability.

Petroleum and flammable fluids are generally classified according to T_F—except for LPG—with further subdivision according to whether the liquids are handled above or below their T_F, as stated in Table 3.1.

It is recommended to apply a subdivision of the Unclassified as for Class III (and Class II) and regard the handling of Unclassified fluids above T_F as Class III (2) fluids.

Above-mentioned classification is often inadequate for a sufficient zone classification under typical processing conditions, which has lead to the introduction of the concept of "fluid category", which is a crucial factor in calculation of the zone hazard radii. Oil fluid categories are listed in Table 3.2.

Note: Class II (1), Class III (1) and Unclassified handled below T_F do not fall into any category unless handled at conditions possible for the creation of a flammable mist, in which case the fluid belongs to category C.

Oil fluids zone classification is touched upon in Chap. 8.

3.2.2 *Limits of Flammability*

Mixtures of flammable gases or vapours in air are only ignitable over a restricted range of concentrations, falling in the range limited by what are termed the Lower Explosive Limit (LEL) and the Upper Explosive Limit (UEL).

The LEL is the minimum concentration of gas or vapor for which flame propagation can occur once ignition has be effected, and the UEL is the maximum concentration that will permit flame propagation following ignition.

Limits of flammability are affected by increasing temperature and pressure—both causing a widening of the range.

LEL is an important parameter for the determination of zone boundaries.

3.2.3 *Auto-Ignition Temperature*

Mixtures of gases or vapours in air will ignite spontaneously when raised to a sufficiently high temperature. The lowest temperature at which this will occur, is termed the auto-ignition temperature (T_{ig}).

T_{ig} values are, as the other flammability characteristics, not absolute constants as they depend very much on the method of measurement. The nature of the surface (e.g. glass contra iron rust) with which the flammable mixture is in contact can affect the value very much.

T_{ig} is an important parameter for the explosion risk assessment and equipment selection—see also Chap. 5, but does not enter zone classification deliberations.

As the temperature equipment classification code for ATEX equipment in Group IIG is also used as a characterizing parameter attached to the respective gas/vapours, this code should be mentioned here.

The max surface temperature of Group IIG equipment is indicated in the Ex-marking by reference to a temperature class (T1 to T6) for use in equipment selection according to the minimum ignition temperature of the gas in question, as stated in Table 3.3.

As an example: A gas having a minimum ignition temperature (T_{ig}) of 305 °C should not get in contact with equipment having a max surface temperature of more

Table 3.3 Temperature class codes

Class code	Max surface temp. (°C)	Suitable when T_{ig} (°C)
T1	450	>450
T2	300	300< and ≤450
T3	200	200< and ≤300
T4	135	135< and ≤200
T5	100	100< and ≤135
T6	85	85< and ≤100

than 300 °C, which on the equipment in the Ex-marking is indicated by the temperature class code “T2”. The gas in question is also said to be a class T2 gas (or a gas belonging to class T2).

Note: Above-mentioned max surface temperature indication of the equipment is supposed to include a safety margin as required by EN 1127-1 and EN 13463-1.

3.2.4 *Gas Group*

Gases and vapours are categorized in terms of their ignition energy or the maximum experimental safe gap (in respect of flameproof protection). This categorization leads to the Gas Groups stated in Table 3.4.

Minimum ignition energies are affected by many factors, and listed values must only be regarded as indications of the order of magnitude. Anyhow, minimum ignition energies for gases and vapours are generally very low—typically 0.2 mJ—and below the values typically encountered in static discharges, so flammable gas/vapour mixtures are always assumed to present an ignition risk. Some gases as ammonia, which has minimum ignition energy of around 680 mJ deviate significantly from the typical value.

Minimum ignition energy is not used in hazardous area classification, but of use in equipment selection (and stated in the Ex-marking).

The low ignition energy Group IIC, exemplified by hydrogen, is the group requiring the most attention.

Apparatus marked for Group IIC can also be used for IIB and IIA gasses, and apparatus marked for Group IIB is also suitable for IIA, but not for IIC.

3.2.5 *Vapor Density*

Vapor density is not a flammability characteristic of gas or vapor, but is an important aspect in hazardous area classification.

Relative vapor density is the ratio of the weight of a volume of a gas or vapor to the weight of an equal volume of air at the same temperature and pressure.

Table 3.4 Gas groups

Group	Representative gases	Ignition energy in air
I	Firedamp (methane) in underground mining	0.2 mJ
IIA	Industrial methane, propane, methanol	High, above 0.2 mJ
IIB	Ethylene, carbon monoxide, ethanol	Medium, between 0.02 and 0.2 mJ
IIC	Hydrogen, acetylene, carbon disulphide	Low, e.g. 0.009 and 0.016 mJ

In hazardous area classification it is important to know how a gas/vapor is likely to move when it is released in order to estimate the shapes and extents of zones, and determining if, and where, layering may occur. The density can contribute to this assessment.

Gases/vapours with relative vapour densities around 1 will diffuse in the air being carried largely by convection currents, etc. Materials with values significantly higher than 1 will tend to settle and layer, and will not mix readily with air. If the gas/vapour has a density significantly less than 1, it will rise and mixing with the air may be slow. Layering will not be a problem, but accumulation at high points may need to be considered.

It is important to note that the release conditions may cause a lower gas temperature than ambient and hence a higher vapour density than the value reported in the literature.

3.2.6 *Summarizing*

The three parameters: Flash point, explosive limits and density, are the important parameters for gas/vapor hazardous area zone classification, which is concerned with whether flammable atmospheres can be formed, their frequency of occurrence and their extents.

3.2.7 *Examples*

Name	Relative density	T_F (°C)	LEL	UEL	T_{ig} (°C)	Gas group
Acetylene	0.9	Gas	2.3	100	305	IIC
Ammonia	0.6	Gas	15	34	630	IIA
Butane	2.1	Gas	1.4	9.3	372	IIA
Carbon monoxide	0.97	Gas	11	74	607	IIB
Ethanol	1.6	12	3.1	19	400	IIB
Hydrogen	0.07	Gas	4	77	560	IIC
Kerosene (diesel No 1)		38–72	0.7	5	210	IIA
Methane (industrial)	0.6	Gas	4.4	17	600	IIA
Methanol	1.1	9	6	36	440	IIA
Petrol	3	−46	1.4	7.6	280	

3.3 Dust Explosion Risk Parameters

The first step is to establish whether the dust is combustible. An explosion classification test can determine whether or not a suspended dust is capable of initiating an explosion and be classified as a Group A dust, which can burn and propagate flames. Group B dusts do not ignite at ambient temperatures, but Group B dusts may present a fire hazard and may be explosible at elevated temperatures (>110 °C).

If the dust is combustible (Group A), the next step is to establish its ignitability.

3.3.1 Physical Properties

The particle size and particle size distribution, particle surface conditions and the composition of the dust, including volatile and moisture content, are important factors.

Where the particles exceed diameters of about 0.5 mm (500 μm), flammable dust clouds will not be formed. Maximum ignitability occurs with particle sizes of less than 75 μm (especially between 10 and 40 μm). Reducing the particle size will reduce (often significantly) the necessary minimum ignition energy (MIE).

Lowering the moisture content will also reduce the MIE, and for example coal dust is generally more highly explosible and exhibit higher explosion violence the greater its volatile content.

3.3.2 Minimum Explosive Concentration (LEL)

Similar to gas/vapor, dust in the form of a cloud is only ignitable over a range of concentrations (i.e. between its limits of flammability), but only the lower limit (LEL) is of practical use, because of the inherent difficulties in achieving homogeneous dust clouds at high concentrations.

The LEL for most dusts is in the range 20–60 g/m^3 (typically about 30 g/m^3), but some materials have values much higher than this, reaching 250–300 g/m^3.

Note: A coal dust cloud of a concentration of 40 g/m^3 will not allow you to see a 25 W bulb at a distance of 2 m.

Temperature and pressure will influence the limits of flammability for dusts in a similar way to those for gas/vapor, but this is of little practical significance as the Upper Explosive Limit (UEL) is not used, and the changes in small concentrations at the LEL will usually have little effects on the hazards.

3.3.3 Minimum Ignition Energy (MIE)

The MIE is a measure of spark sensitivity—used for assessing static and mechanical ignition sources. MIE is affected by many factors including: Plant conditions as temperature, pressure, turbulence and oxygen concentration, powder properties as particle size, exact chemical composition and moisture content, and test conditions as electrode details and discharge circuit.

MIE of dust clouds varies over at least eight orders of magnitude, from about 0.01 mJ for very fine sulphur powder to beyond 1000 J at the upper end.

As mentioned in Sect. 2.4, MIE for vapor/gas mixtures in air are comparatively low in general, varying from 0.01 to 1.0 mJ (typically 0.2 mJ), which are energies below the values typically encountered in static discharges, and so flammable vapor/gas mixtures are assumed always to present an ignition risk.

MIE for dust is often reported to fall between 10 and 100 mJ (or above), but values down around 1 mJ and significantly higher values are also seen in the literature. MIEs for dust are reduced significantly at elevated temperatures and by the presence of flammable gas/vapor, even below the LEL.

MIEs for dust are not used in area zone classification, neither the specification of apparatus. The primary use is in the control of electrostatic hazards (With MIE above 500 mJ, the dust cloud is normally considered as being non-sensitive to ignition by electrostatic discharges).

Where an exact MIE value is of importance for explosion risk assessments, a representative sample of the relevant dust has to be investigated (tested).

In the literature you may find the MIEs (only to be regarded as examples which according to the actual circumstances may deviate considerably), as shown in Table 3.5.

Figure 3.1 conveys some aspects between MIE energies for gases and dusts and possible ignition sources.

Table 3.5 MIE examples from the literature

Dust material	MIE (mJ)
Aluminium	<1
Bio-waste	>1000
Coal	>1000 and 60
Anthracite	100
Bituminous coal	30
Lignite	30
Charcoal	20
Magnesium	>1000 and 80
Zinc	300 and 9600
Polystyrene	100
Urea	100

(continued)

Table 3.5 (continued)

Dust material	MIE (mJ)
Calcium stearate	10
Cornstarch	10 and 30
Corn flour	20 and 40
Pea flour	40 and 100
Sawdust	10
Sugar	10 and 30
Sewage sludge	100 and 8000
Sulphur	<1
Yeast	100
Wood	40

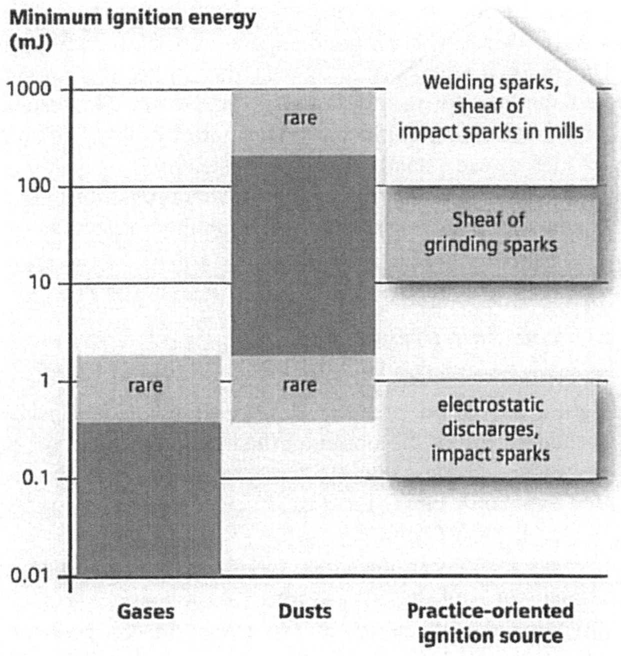

Fig. 3.1 Minimum ignition energy of different materials

3.3.4 Minimum Ignition Temperature (MIT or T_{ig})

MIT (or T_{ig}) is the lowest temperature of a heated surface that ignites a dust cloud upon a brief contact. MIT is dependent on the shape and surface of the vessel used for its determination.

MIT does not inter zone classification deliberations, but is of importance when considering the elimination of ignition sources and for the specification of electrical equipment, and is a design parameter for explosion suppression.

3.3.5 *Layer Ignition Temperature (LIT or T_s)*

LIT of a dust layer is the lowest temperature at which a dust layer on a hot surface ignites. For a dust layer of thickness 5 mm (the standard reference value), the LIT is often referred to as the smoulder temperature (T_s), or sometimes the glow temperature.

LIT shows often a nearly linear decrease as the thickness increases due to the difference in heat loss.

Tests are usually carried out on 5 mm thick layers, as layers in hazardous areas should not reach this depth. The ignition temperatures obtained from the test can be influenced by the thermal prehistory of the material, so that it is important to ensure that the sample submitted for testing is representative (A circumstance of general importance when determining explosion parameters for dust, which means that it may be a better idea to use "standard" values instead).

LIT does not enter zone classification deliberations, but is an important parameter for explosion risk assessments and equipment selection.

3.3.6 *Max Explosion Pressure (P_{max})*

P_{max} is the highest explosion pressure developed by an enclosed explosion—measured in a standard test at the optimum dust concentration.

Note: If a pressure rise of less than 0.5 bar is recorded for all dust concentrations across a potential explosible range, the dust is considered to be non-explosible, and the dust is allocated to dust Group B.

P_{max} which generally falls in the range 6–10 bar, does not enter zone classification deliberations, but is of importance for explosion risk assessments and when considering mitigating the effects of an explosion—in cases where the possible occurrence must be accepted. Examples of dust K_{st} values are listed in Table 3.7.

3.3.7 *K_{st} Value (Max Rate of Pressure Rise)*

The max rate of pressure rise $(dP/dt)_{max}$ is the highest rate of pressure rise generated by an enclosed dust explosion—measured in a standard test at the optimum dust concentration.

Table 3.6 St classes

St class	K_{st} value (bar m s^{-1})	Characteristics
St 0	0	Non-explosible
St 1	$0 < K_{st} \leq 200$	Weak to moderately explosible
St 2	$200 < K_{st} \leq 300$	Strongly explosible
St 3	$300 < K_{st}$	Very strong explosible

Table 3.7 Literature examples of dusts and K_{st} values

Dust	Micron	K_{st}
Aluminum powder	22	400
Brown coal	41	123
Charcoal	29	117
Magnesium	28	508
Paper tissue dust	54	52
Rice starch	18	190
Soap	65	111
Soy bean flour	20	110
Tobacco	49	12
Wood dust	43	102

As $(dP/dt)_{max}$ is volume dependent, a volume-independent dust specific explosibility characteristic K_{st} (called the kst value) is calculated as follows:

$$K_{st} = (dP/dt)_{max} \times V^{1/3} \left(\text{having the unit: bar m s}^{-1}\right),$$

where "V" is the total volume of the vessel (m^3).

The K_{st} value is derived only from measurements in either a 1 m^3 vessel or a 20 litre sphere container.

The K_{st} value is used to classify dusts into several groups or **St classes**, as indicated in Table 3.6.

As P_{max}, the K_{st} value does not enter zone classification deliberations, but is of importance for explosion risk assessments—and when considering mitigating the effects of an explosion—in cases where the possible occurrence must be accepted (Table 3.7).

3.3.8 Powder Volume Resistivity

Powder volume resistivity is not a flammability property, but may be of importance for equipment selection as a conducting layer of dust could cause short-circuits within electrical equipment.

Table 3.8 Some resistivity values

Dust	Resistivity ($\Omega \times m$)
Carbon, amorphous	6×10^{-4}
Carbon graphite (depending on the plane)	4×10^{-6} and 3×10^{-3}
Diamond	10^{12}
Wood	10^{14}–10^{16}

The Powder Volume Resistivity ($\Omega \times m$) is a measure of the resistivity through the bulk of the powder and indicates the ability of the powder to retain charge. It is used to assess the likelihood of accumulating electrostatic charge on the bulk powder, especially when the MIE is low, say below 25 mJ.

The volume resistivity (ρ) in unit "$\Omega \times m$" of a specimen having an area A (m^2) and a length l (m) is determined by the equation:

$$R\ (\text{Resistance}) = \rho \times (l/A)\ \Omega\ (\text{Ohm})$$

Note: 1 $\Omega \times m$ = 100 $\Omega \times cm$

Volume resistivity covers a very broad scale from about 10^{-8} $\Omega \times m$ for metals to about 10^{16} $\Omega \times m$ for super insulators.

In the literature you find for example the resistivity values as indicated in Table 3.8.

EN 60079-20-2 (If/when published) defines, in continuation of the specifications in EN 60079-0, three dust groups (capable of forming explosive atmospheres) with the resistivity as one of the parameters, as stated in Table 3.9.

In North America explosive dust is subdivided into the three material groups indicated in Table 3.10.

Table 3.9 Suggested dust group definitions based on resistivity

Dust group	Type	Definition	Examples
IIIA	Combustible flyings	Solid particles, including fibres, <500 µm Can burn or glow in air	Rayon, cotton, sisal, jute, hemp, oakum, waste kapok
IIIB	Non-conductive combustible dust	Fine solid particles, ≤500 µm Resistivity $>10^3$ $\Omega \times m$	Corn, custard powder, cellulose acetate, nylon, sugar, polythene, epoxy
IIIC	Conductive combustible dust	Fine solid particles, ≤500 µm Resistivity $\leq 10^3$ $\Omega \times m$	Aluminium, magnesium, tin, cadmium, coal dust[a]

[a]Not all types of coal dust are to be regarded as conductive dust

Table 3.10 North American dust groups

Dust group	Specification	Examples
E	Metal dust (explosive and conductive)	Aluminium, magnesium
F[a]	Carbonaceous dust (explosive and some conductive)	Carbon black, coal dust, coke dust
G	Grain dust (explosive and non-conductive)	Flour, sugar, spices, certain polymers

[a]Group F is often specified with the requirement of >8 % entrapped volatiles

Powders with a volume resistivity $<10^9\ \Omega \times \text{m}$ are considered to have low to medium resistivity. It is rare for powders to have resistivity much lower than this. Even metal powders can have significant resistivity as an insulating oxide film builds up on the surface of them.

Note: The designation of the degree of conductivity/resistivity and accompanying value specifications do not seem to be totally international harmonised (Regarding surface resistivity, the relative humidity (RH) of the test sample should always be stated).

The ability of charge to accumulate on powder is dependent on the Powder Volume Resistivity ($\Omega \times \text{m}$) of the powder, and the charge relaxation time, which is a function of Powder Volume Resistivity. In theory, the greater the Powder Volume resistivity, the higher the charge relaxation time will be. Both of these parameters are affected by relative humidity and therefore should be tested accordingly. In theory, the higher the relative humidity, the lower the Powder Volume Resistivity, and the lower the charge relaxation time will be.

Bulk powders are often split into three volume resistivity categories as follows:

Low resistivity	$<10^6\ \Omega \times \text{m}$
Medium resistivity	10^6–$10^{10}\ \Omega \times \text{m}$
High resistivity	$>10^{10}\ \Omega \times \text{m}$

In practice, low resistivity powders are rare because metal powders will eventually oxidize.

When Powder Volume Resistivity is $>10^9\ \Omega \times \text{m}$, there is no electrostatic risk in earthed metal plants provided that Minimum Ignition Energy is >10 mJ.

Equipment "protected by enclosure" relies on excluding the dust from the inside of the equipment to prevent ignition. The degree of protection provided by enclosures are specified by the **IP code** (Ingress Protection code), see also EN 60529.

Equipment of Category 1 and 2 (appropriate for zone 20 and 21, respectively) needs to be built to IP6X specifications (Dust-tight enclosure), and equipment of Category 3 (appropriate for zone 22) in general to IP5X specifications (Dust-protected enclosure). But where the possible explosive atmosphere consists of **conductive dust**, electrical equipment for use in zone 22 must fulfil IP6X specifications. I.e. when ordering Category 3 electrical equipment it must be emphasized that IP6X specifications must be complied with (alternatively, Category 2 electrical equipment may be procured—when economically feasible).

3.3.9 Housekeeping

The housekeeping is of course not a flammability parameter of the dust in question, but outside equipment at the plant, the most important anti-explosion measure—apart from placing the equipment in open air, is to maintain the process buildings in a clean condition. The housekeeping standard can significantly influence the zone classification and the risk of secondary explosions.

3.3.10 Summarizing

The three parameters: Physical properties (size distribution, volatile and moisture content), Lower explosive limit (LEL) and the standard of housekeeping are the important parameters for dust hazard area zone classification These parameters are of special relevance concerning the probable formation of flammable dust atmospheres, the frequency of occurrence and the extent of the hazardous areas.

Note: The very important parameter "Flash point" that is used to indicate the flammability for gas/vapours is not appropriate for dust.

3.3.11 Examples

K_{st} for some dust types (examples)

Dust	K_{st} (bar m s^{-1})
Flour	87
Starch	150
Aspirin	217
Pigment	286
Aluminium	555

Explosion relevant parameters are very much dependent on especially dust size and distribution and moisture content, so the following data should only be used as illustrative examples:

Eksplosibility parameters for some non-coal dusts

Dust type	LEL (g/m^3)	P_{max} (bar)	K_{st} (bar m s^{-1})	T_{ig} (°C)	T_s (°C)
Wood dust	30–60	8.9–10.5	130–190	420–520	310

(continued)

(continued)

Dust type	LEL (g/m^3)	P_{max} (bar)	K_{st} (bar m s^{-1})	T_{ig} (°C)	T_s (°C)
Rice	60	6.7	91	510	450
Wheat flour	60–125	8.0–9.3	139	440	440
Sugar	30	9.0	123	370	400
Cellulose	125	8.9	56	480	270

Eksplosibility parameters for some coal dusts

Coal dust	LEL (g/m^3)	P_{max} (bar)	K_{st} (bar m s^{-1})	T_{ig} (°C)	T_s (°C)
Braun coal	30	8.5	140	430	240
Lignite	60	9.3	110–175	430	240
Bituminous–high volatile	30–60	9.1	70	580	240
Bituminous–low volatile	30–60	8.3	80	690	285
Pet coke	30–125	7.2	45	695	320
Anthracite	100	0.6	2	710	360

LEL Lower explosive limit
P_{max} Max explosion pressure
K_{st} "Explosion constant"—Max rate of pressure rise (see above)
T_{ig} Minimum dust cloud ignition temperature
T_s Minimum dust layer (5 mm) ignition temperature

Chapter 4
ATEX—Prevention

4.1 Introduction

In general, explosion prevention can be achieved by one (or more) of the following actions:

- Avoidance of the fuel (the flammable material)
- Working outside the flammable range
- Removal or reduction of oxygen content (inerting)
- Eliminating all ignition sources

The first three actions will prevent the formation of an explosive atmosphere, which has first priority amongst the basic principles in explosion prevention and protection.

4.2 Avoidance of Fuel (Flammable Materials)

Solving explosion risk problems by substituting flammable materials with non-flammable materials is of course an obvious initiative, but impossible in many practical cases—or at least very difficult.

Concerning flammable liquids, one should aim at operating at least 5 °C below the Flash point (T_F) in order to have a reasonable safety margin.

The risk of mist explosions can be minimized by operating at lower pressures and temperatures, using more viscous liquids, and of course by avoiding leaks.

Especially for dust, the use of larger particle sizes (>0.5 mm) minimizes the explosion risks, and the addition of an inert substance to the dust can prevent the formation of an explosive atmosphere.

Inert dusts, as calcium sulphate, limestone, sodium bicarbonate, common salt, various silicates or stone dust may be used, but the method requires in most cases at

T. Jespen, *ATEX—Explosive Atmospheres*, Springer Series in Reliability Engineering, DOI 10.1007/978-3-319-31367-2_4

least the addition of 60 % diluent dust, and an intimately mixing of the two dust components. Outside the coal mining field, this prevention method has a very limited use.

In certain cases, the addition of water or a high flash point liquid to a flammable dust is also a possibility for avoiding an explosive dust atmosphere.

Outside equipment, a good housekeeping standard can be very important for avoiding secondary dust explosions (i.e. removal of the fuel, which could result in an explosive dust cloud).

Explosions are very uncommon in open areas, therefore equipment that may create a possible explosive atmosphere in their surroundings, should never be located inside closed buildings or similar structures.

4.3 Keeping Conditions Outside the Flammable Range

When flammable gas/vapour or dust is kept outside the explosive range, i.e. outside the concentration range between LEL (Lower Explosive Limit) and UEL (Upper Explosive Limit), the atmosphere is not explosive. But for many processes this solution is without practical importance, or the control measures for ensuring the process to be below or above explosive limits, are too unreliable.

Outside equipment, **ventilation** can be an appropriate way of avoiding the flammable range (keeping the fuel concentration below LEL) for gases and vapours, and minimize the extent of a hazardous zone.

General ventilation is of limited use concerning dust, but appropriate point source ventilation (dust exhaust) can contribute effectively to minimize the creation of explosive dust atmospheres outside equipment.

Where ventilation is used, care must be taken to ensure that it is not possible to develop flammable concentrations in the ventilation system itself, say due to deposits of dust forming at changes in cross-sections or on the inside of bends.

4.4 Inerting

When the explosion risks cannot be eliminated by material (fuel) substitutions or adjustment of process condition to ensure substances being outside the flammable range, inerting comes into focus.

The partial or complete substitution of the air (oxygen) by an inert gas is a very effective method in explosion prevention.

Every flammable material (gas, vapor, mist or dust) has associated with it a parameter known as the Limiting Oxygen Concentration (LOC) valid for the inert gas in question. LOC is the oxygen concentration below which—regardless of the concentration of flammable material—flame propagation and consequently an explosion, cannot occur.

With lignite (brown coal) for example, the LOC amounts to approximately 12 % by volume using nitrogen (N_2) as inert gas, and approximately 14 % when using carbon dioxide (CO_2).

In practice, the derived maximum allowed O_2 concentration (MAOC) is an operational parameter that is set approximately 2–4 % points below LOC (e.g. with an LOC of 12 %, the MAOC will be set no higher than 10 %).

Note: Being below the LOC is not sufficient for the extinguishing of for example a smouldering coal dust fire which is only possible at an O_2 concentration as low as 2–3 %.

In general, LOC values for solvents and gases are in the range 8–10 % (v/v), whilst for dust they fall in the range 10–14 % (v/v).

A major risk associated with the use of inerting is that of asphyxiation, particularly in confined spaces, which therefore requires appropriate precautions and control measures in those events where people are required to enter these areas.

Note: Inerting is of course only of practical relevance regarding processes which are not substantially open to the atmosphere.

Important inert (as far as concerning fire circumstance) gases, according to their effectiveness, are:

1. Carbon dioxide (CO_2)
2. Steam (H_2O)
3. Flue gases (e.g. from the cement rotary kiln or hot gases from a hot gas generator which contain low O_2 levels)
4. Nitrogen (N_2)
5. Noble gases such as Argon and Helium

Information on inerting can be found in the guidelines:

- CEN/TR 15281
- VDI 2263-2
- BGV C15

Inerting systems are not regarded as protective systems in the meaning of the ATEX Equipment directive 2014/34. If the inerting system is to be placed in an ATEX zone classified area, it must of course comply with the relevant requirements of the directive.

A conservative guideline regarding the practical significance of inerting is saying that inerting leads to a one step down in the zone classification, i.e. you move from a zone 20 to a zone 21, or from a zone 21 to a zone 22, because the intended inert conditions cannot be guaranteed all the time.

The practice of inerting is also employed in explosion suppression systems (see Chap. 6), where typically a quick acting pressure switch responds to the initial comparatively slow increase in pressure due to initiation of explosion conditions. This in turn triggers injection of an explosion suppressant such as chlorobromomethane, water or carbon dioxide into the path of the advancing flame front.

4.5 Carbon Dioxide (CO_2)

CO_2 is a colourless, odourless, non-corrosive and electrically non-conductive gas with a density approximately 50 % greater than air. It is generally stored in the liquid phase, under pressure.

CO_2 is a minor (but important) constituent of the atmosphere, averaging about 0.03 % ($\sim$300 ppm by volume). Since dry CO_2 is a relatively inert gas, special materials of construction are not required, but where moisture may be present in high concentrations, carbonic acid can be formed, and materials resistant to acids should be used.

Some carbon dioxide properties are listed in Table 4.1.

Bulk CO_2 is typically stored as a liquid in storage tanks with capacities of approximately 6, 14, 26 and 50 tons (capacities of 90 tons have also been mentioned), maintained at a pressure between 17 and 21 bar by a refrigeration unit (Low pressure systems).

Smaller liquid quantities are stored and shipped in cryogenic liquid cylinders with a capacity of about 175 kg. The vapour pressure of the liquid in the cylinders at 20 °C is about 56 bar (High pressure systems).

High pressure storage CO_2 systems are also available with tank sizes from 3 to 15 tons which are supplied from low pressure tankers with a max pressure between 20 and 25 bar of CO_2 being kept at a temperature between −20 and −29 °C.

4.5.1 CO_2 Inerting at Coal Grinding

Inerting is an important matter in avoiding explosions and smouldering fires in silos, mills and filter equipment.

In normal operation at cement plants inerting is created with the exhaust gases from the rotary kiln or from a hot gas generator during the operation of the coal mill department. But in case of an emergency shutdown or at the starting and stopping of the coal mill, the injection of inert gas is needed.

The carbon dioxide density tells us that 2 kg of liquid CO_2 is necessary for the supply of 1 m^3 of inert gas. According to the German coal dust regulation BGV C15, a sufficient inertisation is guaranteed if the geometrical (empty) volumes of the

Table 4.1 Carbon dioxide properties

Molecular weight	44 kg/kmol
Density at 0 °C and 1 bar	1.98 kg/m^3
Relative density (to air)	1.5
Triple point	−56.6 °C, 5.2 bar
Critical point	31.0 °C, 73.8 bar
Pressure at −18 °C	20.7 bar
Pressure at +21 °C	58.6 bar

individual components receive a simple flushing, i.e. in a 1:1 ratio (It is not absolutely necessary to replace all the oxygen by inert gas).

But extinguishing smouldering fires requires an oxygen concentration less than 2–3 %, which means that the inerting process has to be repeated up to 3 or 4 times, depending on the LOC when inerting was started.

In the flushing inerting method, carbon dioxide is introduced at the highest possible speed (avoiding the activation of any explosion vents present) into different areas of the system to be inerted, which produces a strong turbulence, thorough mixing of the gases and an optimum inerting, because pockets of high oxygen concentration are avoided.

Note: Liquid CO_2 cannot be discharged directly into the coal, because the expansion of the liquid to atmospheric pressure would form small particles of dry ice which will plug any application points buried in the coal. CO_2 have to be used as a gas, generally below 2 bar.

Yara (a Norwegian company offering inerting systems and gases) has recommended the following design criteria: Storage of 2–3 times the maximum necessary inert gas volume (see above) supplemented by a security reserve, and the possibility of withdrawal the max amount of inert gas within one hour.

High-Pressure CO_2 Inerting Systems

High pressure systems use tanks with capacities of 3–15 tons CO_2 or a battery consisting of standard CO_2 steel cylinders, with the storage of CO_2 at ambient temperatures (max 25 °C), and pressures in the range from 50 to 70 bar.

Standard steel cylinders are used for small to medium CO_2 inerting capacities in places with limited infrastructure and where CO_2 is not available by road.

The tanks with higher capacities are always filled from low pressure tankers (max pressure of 20–25 bar), supplying the CO_2 cooled down to a temperature between −20 and −29 °C.

The high pressure tanks may have to be cooled during the summer period either with cooling water or by being contained within an air-conditioned room.

In order to keep the tanks at the operating pressure of 50–70 bar during winter periods, they are equipped with heaters.

High pressure systems are mainly used in countries with large seasonal temperature fluctuations, such as those in Europe, Russia, central Asia and parts of the Americas. Maybe because a low pressure system, which is always equipped with a cooling system, is preferred when cooling is required during long periods of the year.

A high pressure system has the general advantage of being independent of electrical power for refrigeration.

Low-Pressure CO_2 Inerting Systems Concerning functionality, these systems are comparable with high pressure systems. The main difference is the integrated refrigeration unit which continuously keeps the liquid at a temperature around −18 °C, and with it at a pressure around 21 bar. For max inert gas discharge, the tanks are equipped with heating equipment to compensate for the drop in pressure.

The CO_2 is taken from the vessels in liquid form, and has to pass through an evaporator to be transformed into the gaseous state. The most suitable kind of evaporators is ambient or atmospheric type because these are independent from any artificial power shutdown.

Low-pressure inerting systems are mainly used in countries with constant temperatures above 5 °C, for example the Middle East, Africa, South Asia, Australia and Central America.

The inert gas delivery capacity is mainly dependent on the size of the evaporator (which ensures that the pressure drop does not lead to too low temperatures) and the ambient temperature.

Ignoring the continuous electricity requirement for refrigeration, it seems reasonable clear that the low pressure system in general is a better choice when the hazards to be tackled are not of a small size scattered throughout a facility.

4.6 Nitrogen (N_2)

N_2 makes up 78 % of the atmosphere, does not support combustion, and at standard conditions is a colourless, odourless, tasteless, non-irritating, and inert gas.

Some nitrogen properties are listed in Table 4.2.

Nitrogen is typically stored and used in equipment at pressures from 0.7 to 207 bar (sometimes as high as 690 bar).

At atmospheric pressure, one volume of liquid nitrogen at its boiling point will vaporize to roughly 700 volumes of gas when warmed to room temperature. Thus, a small liquid nitrogen leak can rapidly displace the surrounding air and create an oxygen-deficient atmosphere.

High pressure N_2 pack inerting systems are used in countries with infrastructure that severely limits the availability of CO_2 by road.

N_2 high pressure packs are normally provided with 9–12 standard 200 bar N_2 steel cylinders for inert gas discharge. Pressure fluctuations caused by climate conditions do not play a role therefore erection outside with weather protection roof as a general rule is sufficient.

Table 4.2 Nitrogen properties

Molecular weight	28 kg/kmol
Boiling point (1 atm)	−195.8 °C
Freezing point (1 atm)	−210 °C
Critical point	−146.9 °C, 33.5 atm
Gas density (20 °C, 1 atm)	1.6 kg/m^3
Relative density (to air)	0.967

Chapter 5
ATEX—Equipment Selection

5.1 Introduction

For areas where a hazardous explosive atmosphere may occur, and therefore require zone classification, the employer must observe the **selection criteria** for equipment and protective systems as defined in the ATEX Workplace directive 1999/92. It means that the selection must be based on the categorisation set out in ATEX Equipment directive 2014/34, **unless** the actual risk assessment (as demonstrated in the Explosion protection document) points at another approach (course of action).

Therefore, without risk assessment based deflections, the combinations of zones and categories stated in Table 5.1 must be observed.

Note: The fundamental ATEX electrical equipment standard EN 60079-0 (and prEN ISO 80079-36 concerning mechanical equipment) contains specifications that categorize the equipment in a slightly different way, using the concept "Equipment Protection Level" (EPL) as a basic parameter—see Sect. 5.6 on "ATEX marking".

Note: The ATEX Workplace directive 1999/92 with its zone classifications does not cover the mining area (where Group I, category M 1 or M 2 equipment—as defined in ATEX Equipment directive 2014/34—must be used). Mineral extracting industries are covered by the Workplace directives 92/91 or 92/104.

Note: Mechanical equipment without own ignition sources are in principle not covered by the ATEX Equipment directive 2014/34. Therefore, such equipment cannot be accompanied by an ATEX EU Declaration of conformity, but may of course be used or installed in zone classified areas.

Note: Protective systems (as defined by the ATEX Equipment directive 2014/34), separately placed on the marked for use as autonomous systems, are not grouped and categorizes as the equipment, which means that the abovementioned zone-category selection scheme is not valid concerning those systems.

As mentioned above, the actual risk assessment may point to the fact that it is appropriate to strengthen the level of protection against own ignition sources of the relevant equipment for the zone in question. As an example, by requiring as

T. Jespen, *ATEX—Explosive Atmospheres*, Springer Series
in Reliability Engineering, DOI 10.1007/978-3-319-31367-2_5

Table 5.1 Zone and equipment relationship

Zone	Equipment category
Zone 0	Category 1 G*
Zone 20	Category 1 D**
Zone 1	Category 1 or 2 G*
Zone 21	Category 1 or 2 D**
Zone 2	Category 1, 2 or 3 G*
Zone 22	Category 1, 2 or 3 D**

G*: Suitable for gas, vapours or mist
D**: Suitable for dust

minimum the use of Category 2 equipment for a Zone 2 or Zone 22, due to anticipated overall direct and indirect effects of an explosion—or vice versa. Moreover, as the ATEX Workplace directive 1999/92 is a minimum directive, the corresponding national legislation may lay down more stringent requirements or supplementary requirements concerning equipment selection.

Besides the basic selection criteria founded on the zone classification of the area in combination with the appropriateness for gas, vapour, mist or dust (identified in the equipment marking by G, D or DG), equipment selection normally has to take into account **supplementary aspects** as temperature conditions, gas or dust group classification, etc. The important supplementary aspects at equipment selection are touched upon in the subsequent parts.

When all relevant ATEX information about the workplace, the process conditions, including processed materials, and the operating requirements of the equipment in question, are available, equipment selection can be carried out based on the equipment marking. The marking must contain all information essential for the safe use of the equipment—if necessary supported by the information in the EU Declaration of conformity—see also Sects. 5.6 and 5.7, and Chap. 12.

5.2 Temperature Considerations

A "hot surface" is one of the very important possible ignition sources. Therefore, the max surface temperature of the equipment (or protective system) is an essential parameter at equipment selection, and must be regarded as an obligatory piece of information in the Ex-marking of ATEX equipment.

The max surface temperature (indicated by temperature class or in °C or both) marked on equipment for use in gas/vapor/mist atmospheres must include a **safety margin** to the minimum ignition temperature (T_{ig}) in accordance with EN 1127-1 and EN 13463-1. Equipment selection (concerning temperature aspects) can be guided by the Table 5.2, which includes the North American code.

Table 5.2 Max surface temperatures and codes (Group II G, category 1, 2 and 3)

Max surface temp. (Safety margin incl.) (°C)	ATEX code	American code	Recommended when T_{ig} (°C)
450	T1	T1	>450
300	T2	T2	>300
280	T2	T2A	>280[a]
260	T2	T2B	>260[a]
230	T2	T2C	>240[a]
215	T2	T2D	>215[a]
200	T3	T3	>200
180	T3	T3A	>180[a]
165	T3	T3B	>165[a]
160	T3	T3C	>160[a]
135	T4	T4	>135
120	T4	T4A	>120[a]
100	T5	T5	>100
85	T6	T6	>85

[a]The ATEX code must then be supplemented by the max temperature value

As an example: For a gas having T_{ig} = 220 °C, equipment with max surface temperature of 215 °C (indicated by ATEX code T2 supplemented by this temperature specification, this temperature specification alone or by the American T2D code) can be selected, but of course, equipment classified in a higher temperature class (e.g. T3 or T4) may also be selected.

For equipment having a max surface temperature greater than 450 °C, only the max temperature value shall be specified in the marking.

Equipment of Group II G, designed and marked for use in a particular gas, does not need temperature specifications.

Dust equipment (Group II D, category 1, 2 and 3) Max surface temperature of equipment for use in dust atmospheres is generally not indicated by the use of temperature classes, but by stating the actual max temperature value.

At equipment selection, the indicated/marked max temperature (T_{max}) must therefore be compared with the dust cloud minimum ignition temperature (T_{ig}) and/or the dust layer minimum ignition temperature (indicated by T_{lig} or T_s), both having been adjusted by an appropriate **safety margin**.

Where dust cannot settle, and ignition by smouldering can be excluded, the max equipment surface temperature (T_{max}) must comply with the equation:

$$T_{max} \leq T_{adig} = 2/3 \times T_{ig}(^{\circ}C), \quad (5.1)$$

where T_{adig} is the adjusted minimum dust cloud ignition temperature.

Where dust can settle, abovementioned criteria must be supplemented by a comparison of T_{max} with an adjusted dust layer minimum ignition temperature (T_{adlig}), which for a max 5 mm dust layer requires that:

$$T_{max} \leq T_{adlig} = T_{lig,5mm} - 75(^{\circ}C), \quad (5.2)$$

where $T_{lig,5mm}$ is the minimum dust layer of 5 mm ignition temperature.

The equipment T_{max} must comply with both Eqs. (5.1 and 5.2) where dust layers can be deposited. If the dust layer thickness may be greater than 5 mm, a larger safety margin is required, as the ignition temperature of dust layers decreases with increasing thickness.

Note: Where the equipment max surface temperature depends not on the equipment itself, but mainly on operating conditions, the equipment is to be marked with "TX", which indicates that the relevant max temperature cannot be marked by the manufacturer, and that the instructions contain more information on this question.

5.3 Gas and Dust Groups

Gases and vapors are classified into explosion gas groups (IIA, IIB and IIC), where the dangerousness increases from gas group IIA to IIC (see also Chap. 12), see Table 5.3.

Electrical equipment approved for group IIC may also be used for the two other explosion groups.

The marking of electrical apparatus (Group II equipment) normally shows for which group it is designed, which can be helpful at equipment selection.

Explosion dust groups The draft EN 60079-20-2 and North American standards also define explosion groups of dust, see Sect. 3.3.8 of Chap. 3, which are particularly based on a differentiation related to the resistivity (or the conductivity) of the dust.

At present, the use of dust groups in equipment selection is probably of limited value, but considerations regarding resistivity of the dust and the proper enclosure protection of electrical equipment may be very important—see subsequent part.

Table 5.3 Explosion gas groups

Gas group	Examples
IIA	Industrial methane, propane, methanol
IIB	Ethylene, carbon monoxide, ethanol
IIC	Hydrogen, acetylene, carbon disulphide

Classification criteria can be studied in EN 60079-1A and EN 60079-3

5.4 Ingress Protection

EN 60529 specifies a classification system for the sealing effectiveness of enclosures of electrical equipment against the intrusion of foreign bodies, as dust and moisture. The system uses the letters **IP** (Ingress Protection) followed by two digits (An "X" is used for one of the digits if there is only one class of protection).

The **First Digit** (0, 1, 2, 3, 4, 5 or 6) indicates the protection against intrusion of foreign objects/dust, and the **Second Digit** (0, 1, 2, 3, 4, 5, 6, 7 or 8) indicates the protection against water intrusion, see Table 5.4.

US uses a classification system set out in NEMA (National Electrical Manufacturing Association) Standard Publ. No. 250. These degrees of protection cannot be equated exactly with those of IEC because NEMA takes account of additional environmental influences (e.g. coolants, cutting coolants, corrosion, icing, hail). Table 5.5 must therefore only be regarded as a non-binding guideline.

Note: Other deviating guidelines concerning comparisons between NEMA and EN can be found in the literature.

Electrical equipment "protected by enclosure" relies on excluding the dust from the inside of the equipment in order to prevent ignition.

Electrical equipment of Group II, category 1 and 2, needs to be built to IP6X specifications (Dust-tight enclosure), and electrical equipment of Group II, category 3, in general to IP5X specifications (Dust-protected). But where the possible explosive atmosphere consists of **conductive dust**, electrical equipment for use in zone 22 (Group II, Category 3 D equipment) must fulfill IP6X specifications (alternatively, group II, Category 1 or 2 equipment, may be procured—when economically feasible).

Table 5.4 IP groups

1st digit	Solid bodies protection	2nd digit	Water protection
0	No protection	0	No protection
1	Objects > 50 mm	1	Drops of water
2	Objects > 12 mm	2	Vertically dripping water
3	Objects > 2.5 mm	3	Sprayed water
4	Objects > 1 mm	4	Splashed water
5	Dust-protected	5	Low-pressure water jets
6	Dust-tight	6	Powerful water jets
		7	Effects of immersion
		8	Continuous immersion

Table 5.5 Best guidance to EN and US IP similarity

Degree of protection NEMA	Degree of protection EN
1	IP10
2	IP11
3	IP54
3R	IP14
3S	IP54
4 and 4X	IP56
5	IP52
6 and 6P	IP67
12 and 12 K	IP52
13	IP54

5.5 Method of Protection

5.5.1 Electrical Equipment

A number of methods of explosion protection for **electrical equipment** have been established and codified in construction standards, see Tables 5.6 and 5.7.

Remarks on the different types of protection:

"i" Intrinsic safety Basic principle: By limiting the energy (current and voltage) in the circuit, the formation of impermissible high temperatures, sparks, or arcs is prevented. The two main subtypes are "ia" (allows for the occurrence of two faults during operation) and "ib" (allows for the occurrence of one fault during operation).

- Covered equipment categories: 1 G, 2 G, 3 G, 1 D, 2 D and 3 D.

"e" Increased safety Basic principle: Precautions are applied to the installation to ensure increased security against the possibility of excessive temperatures and sparks from the equipment. Equipment that normally causes sparks is excluded.

- Covered equipment categories: 1 G and 2 G.

"d" Flameproof Basic principle: Equipment that may cause an explosion is contained within an enclosure which can withstand the force of an explosion and prevent transmission to the outside hazardous atmosphere.

- Covered equipment categories: 2 G and 3 G.

"p" and "pD" Pressurised Basic principle: The "p" type protection consists in introducing an ignition shield gas (composed of air or inert gas as nitrogen) into the

Table 5.6 Electrical equipment types of protection concerning gas

Protection type	Ex code	Zones	Comments
Intrinsically safe (EN 60079-11)	Ex "i"	Zone 0: "ia" Zone 1: "ib" Zone 2: "ic"	For instrumentation, control and measurements
Increased safety (EN 60079-7)	Ex "e"	Zone 1 or 2	For lighting, junction boxes and motors
Flameproof (EN 60079-1)	Ex "d"	Zone 1 or 2	For lighting, junction boxes, motors, electronics
Pressurised/purge (EN 60079-2)	Ex "p"	Zone 1: "px" or "py" Zone 2: "pz"	For analysers, motors, control boxes, computers
Powder filling (EN 60079-5)	Ex "q"	Zone 1 or 2	For electronics, telephones, chokes
Encapsulation (EN 60079-18)	Ex "m"	Zone 0: "ma" Zone 1: "mb" Zone 2: "mc"	For electronics
Oil emission (EN 60079-6)	Ex "o"	Zone 1 or 2	For heavy current equipment
Type n (non-incendive) (EN 60079-15)	Ex "n"	Zone 2	For lighting, motors, junction boxes, electronics
Special protection (TR 60079-33)	Ex "s"	Zone depending according to certification	According to certification

Table 5.7 Electrical equipment types of protection concerning dust

Protection type	Ex code	Zones	Comments
Enclosure (EN 60079-31)	Ex "tD"	"tDA" with IP6X: Zone 21 and 22 "tDA" with IP5X: only Zone 22 with non-conductive dust	A method "B" can also be used A ident.: tDA20, tDA21, tDA22 B ident.: tDb20, tDB21, tDb22
Pressurised enclosure (EN 60079-2)	Ex "pD"	Zone 21 or 22	
Intrinsic safety (EN 60079-11)	Ex "iD"	Zone 20: "iaD" Zone 21: "ibD" Zone 22: "icD"	
Encapsulation (EN 60079-18)	Ex "mD"	Zone 20: "maD" Zone 21: "mbD"	

enclosure in order to keep it at a pressure above that of the external atmosphere, thus preventing the formation of an explosive atmosphere inside the enclosure itself.

The internal overpressure technique may also be applied, essentially with the same characteristics, to equipment for dust; in this case the marking is “pD”.

- Covered equipment categories: 2 G, 3 G, 2 D and 3 D.

“q” Powder filling Basic principle: All equipment that has the potential to arc is contained within an enclosure filled with a material in a finely granulated state (e.g. quartz or glass powder particles). The powder filling prevents the possibility of an ignition.

- Covered equipment categories: 2 G and 3 G.

“o” Oil emission Basic principle: Equipment or equipment parts are immersed in oil to the effect that an explosive atmosphere above the surface or outside the enclosure cannot ignite.

- Covered equipment categories: 2 G and 3 G.

“m” Encapsulation Basic principle: Parts which may ignite an explosive atmosphere are embedded into a sealing compound with sufficient resistance to guards against an explosive atmosphere, sparks or heat (i.e. segregating/encapsulating possible ignition sources).

- Covered equipment categories: With “m” marking: 1 G, 2 G and 3 G, and with “mD” marking: 1 D, 2 D and 3 D.

“n” Type n protection (non-igniting) Basic principle: Protection method(s) for electrical equipment designed so that it will not ignite the surrounding explosive atmosphere in normal operation and under certain fault conditions specified in the standard. The “n” type covers 5 different categories: nA, nC, nR, nL and nP.

- Covered equipment category: 3 G.

“tD” Enclosure Basic principle: Ignition protection is based on the limitation of the maximum surface temperature of the enclosure and other surfaces which may come into contact with dust, and on the restriction of dust ingress into the enclosure by the use of “dust-tight” (IP6X) or “dust-protected” (IP5X) enclosures.

- Covered equipment categories: 1 D, 2 D and 3 D.

5.5.2 *Non-electrical Equipment Types of Protection*

Opposite electrical equipment, which are generally regarded as a potential ignition source always needing special protection, mechanical (non-electrical) equipment intended for use in ATEX zone classified areas must first pass an **Ignition hazard assessment**, e.g. according to the specifications of EN 13463-1—see Chap. 12.

The Ignition hazard assessment shall reveal the presence of any equipment own possible ignition sources, and where ascertained, establish the likelihood of them to become an effective ignition source in order to categorize the equipment.

Mechanical equipment without own ignition sources are not covered by ATEX equipment directive 2014/34 and can be used in ATEX zones without fulfilling specific ATEX equipment requirements.

EN 13463-1 is the basic standard concerning non-electrical equipment for use in ATEX environments specifying the basic requirements to be fulfilled by the equipment of all categories of Group I and II (Most category 3 mechanical equipment needs only to meet the requirements put down by EN 13463-1).

Where supplementary protection is needed other standards of the EN 13463 series can be relevant:

EN 13463-2	Protection by flow restricting enclosure (Code: fr)
EN 13463-3	Protection by flameproof enclosure (Code: d)
EN 13463-5	Protection by constructional safety (Code: c)
EN 13463-6	Protection by control of ignition sources (Code: b)
EN 13463-8	Protection by liquid immersion (Code: k)

At the moment, the EN standards are being used as a basis for developing international standards:

- An ISO 80079-36 standard on basic methods and requirements in line with EN 13463-1
- An ISO 80079-37 standard with types of protection “c”, “b” and “k”, based on 13463-5, -6 and -8.

5.6 ATEX Marking

Marking requirements (Ex-marking) and marking recommendations are dealt with in more detail in Chaps. 12 and 16, which sets focus on the ATEX Equipment directive 2014/34, dealing with manufacturer obligations e.g. concerning placing on the market and putting into service of ATEX equipment.

The ATEX Ex-marking shall in principle provide all the necessary information for an appropriate and correct choice of equipment, and for its safe use or

Table 5.8 Marking according to EN 60079-0 and ATEX directive

EN 60079-0		ATEX directive 2014/34		
EPL	Equipment group	Equipment group	Category	Zone
Ma	I	I	M 1	NA
Mb			M 2	
Ga	II	II	1 G	0
Gb			2 G	1
Gc			3 G	2
Da	III		1 D	20
Db			2 D	21
Dc			3 D	22

installation in ATEX zones **but** the markings you experience in practice may be rather confusing. The reason is that they are not really "standardised" and the recommendations in relevant standards are not aligned with the classification scheme of the ATEX Equipment directive 2014/34.

In any case, the specific equipment marking must be explained in the accompanying user instructions.

The fundamental ATEX standard, EN 60079-0, on electrical equipment contains specifications that categorize the equipment in a slightly different way compared to ATEX directive 2014/34, using the basic parameter "Equipment Protection Level" (EPL), see Table 5.8.

Virtually, the EN 60079-0 classification differs only from the ATEX directive classification regarding one aspect: The more distinct splitting of the two types of explosive atmospheres (gas, vapour and mist on one side, and dust on the other) into the two Groups II and III.

The CE-/Ex-marking on ATEX equipment and protective systems normally (especially on electrical equipment) consist of the following particulars in the order listed here:

- CE-marking accompanied by the identification number of the Notified Body involved in the production control phase **when** the involvement of a Notified Body is foreseen by the directive (voluntary certification should not lead to the affixing of the number of the Notified Body, in order to avoid confusion).
- The specific explosion protection marking: "εx" inside a hexagon.
- Equipment Group according ATEX directive (I or II) or EN 60079-0 (I, II or III).
- Equipment Category (1, 2 or 3).
- For equipment Group II (according to ATEX directive) or Group II and III (according to En 60079-0): G, D or DG according to relevant explosive atmosphere suitability.
- Information on explosion protection starting with "Ex".

- Specification of the type of protection using the Ex code, e.g. "tD A21" (see Table 5.7). More than one type of protection may be indicated in the marking.
- Specification of Gas or Dust group, e.g. IIC, as applicable.
- Specification of Ingress Protection (IP), e.g. IP66, as applicable.
- Specification of max surface temperature, either by temperature class, e.g. T4 (for gases) and/or temperature in °C, e.g. "T80 °C" (for dust).
- Finally, the EPL level may be specified, e.g. "Gb".

If all the possible marking elements are use, some information duplication most certainly will occur.

In some cases an ATEX product is not meant to be surrounded by only one specific zone. One part (e.g. the interface to the process flow) belongs to one category (e.g. 1 D) and another part (e.g. the interface to the external surroundings) belongs to another category (e.g. 3 D) or to no category (where this part is not intended for use in a potential explosive atmosphere).

Where more than one categorization is relevant a "/" is used between the two identified relevant categories, e.g. 1/3, and where some part is not to be placed in an ATEX zone, to be indicated by "-", e.g. 1/-.

Safety devices, controlling devices and regulating devices (Article 1.2 devices), intended for use outside ATEX zones, but required for or contributing to the safe functioning of equipment and protective systems with respect to explosion risks, also fall under the scope of the ATEX equipment directive 2014/34.

Article 1.2 devices, separately put on the market are to be CE-/Ex-marked, indicating the category of the equipment to be under their control in round brackets. As an example, the marking "II (1) G D" would cover an Article 1.2 device to be placed outside ATEX zones, which can be connected e.g. to Category 1 equipment. The marking "II 2(1) G D" covers a Category 2 equipment containing an Article 1.2 device for a Category 1 equipment.

Components, in ATEX equipment directive 2014/34 defined as "any item essential to the safe functioning of equipment and protective systems but with no autonomous function" must also comply with directive 2014/34 obligations, but components must not have CE-marking affixed (unless required by other directives).

Although no CE-marking is stated in the directive, expert recommendations for the opposite are also encountered.

Components are required to be accompanied by Attestations of conformity, see subsequent Sect. 5.7.

An **Assembly**, formed by combining two or more pieces of ATEX equipment, is a product under the scope of the ATEX Equipment directive 2014/34, provided that this assembly is placed on the market and/or put into service as a single functional unit. Such assemblies will often require proper installation at the place of use.

When a final combination of equipment (e.g. some ATEX mechanical equipment which needs to be connected to different pieces of ATEX electrical equipment) does not create additional ignition hazards (determined on the basis of an ignition risk assessment), no further action is required. The compliance of the assembly is then based on the compliance of the individual parts.

The assembly manufacturer may also regard the assembly as a piece of ATEX equipment, affix CE- and Ex-marking, sign a EU Declaration of conformity, and provide instructions for the safe use of the assembly. In doing so, the assembly manufacturer takes complete responsibility, and this procedure does not require any renewed involvement of a Notified Body.

Where an assembly consists of parts not already in full conformity with relevant parts of the Directive, or the ignition risk assessment has revealed additional/new risks, the assembly must undergo the complete conformity assessment procedure appropriate for the category.

ATEX equipment Ex-marking examples

- II 2 G Ex d IIC T4 Gb:
  Group II, Category 2, Gas atmosphere equipment, type of protection "d" (Flameproof), Gas group "IIC", temperature class T4, EPL level "Gb" (ATEX directive: II 2 G).
- III 2 D Ex tD A21 IP66 T80 °C Db:
  Group III, Category 2, Dust atmosphere equipment, type of protection "tD A21" (Enclosure), Ingress protection class "IP66", Max surface temperature "80 °C". EPL level "Db" (ATEX directive: II 2 D).
- Ex ch + kh IIB T4 Ga:
  Protection using a combinations of two independent types "ch" and "kh" to obtain an EPL level "Ga", equipment group II (ATEX directive: II 1 G), Gas group IIB, temperature class T4.

5.7 ATEX EU Declaration and Attestation of Conformity

More information on ATEX EU Declarations and Attestations are to be found in Chap. 12.

The accompanying **EU Declaration of conformity** must state all relevant provisions fulfilled by the equipment, protective system or Article 1.2 device, which should be covered by the Ex-making and be repeated here, e.g. Equipment Group II, Category 2 G (IIB T4).

Note: Simply to declaring a product as being in conformity with the ATEX Equipment directive 2014/34 as such (in line with the general statement in a Machinery Declaration of conformity) is inadequate, because no product can comply with all the obligations at the same time, and because the Directive requires the relevant provisions fulfilled to be specified. Directive 2014/34 put forward obligations to be fulfilled according to the relevant product (equipment, protective system or device), Group and Category, and the relevant specific compliance must appear from the Ex-marking and the EU Declaration of conformity.

The **Attestation of conformity**, which must accompany components, as mentioned above, shall declare the conformity with the provision of Directive 2014/34 which apply to them, state their characteristics and how they are to be incorporated into equipment or protective systems.

As for the Declaration, the conformity declaration specification in Attestations could take the form of a relevant Ex-marking (should that have been required by the Directive), e.g. "Ex II 1/2 GD cb TX" or "Ex II 1 GD c TX".

Chapter 6
ATEX—Mitigating Measures

6.1 Introduction

When you cannot prevent the formation of an explosive atmosphere, and you cannot avoid the possible presence of ignition sources, e.g. because they may be introduced as elements of the process flow, the possibility of an explosion must be accepted.

In order to ensure the safety of employees, mitigating measures then have to be decided upon and implemented.

Mitigating measures can be grouped as:

- **Explosion relief venting**—based on the opening of a specifically designed weaker area (membrane) of the equipment in order to discharge burnt and un-burnt material into an appropriate safe location (normally into open air) during the incipient stages of an explosion, and thereby limiting the internal explosion pressure to a sufficient safe level for the equipment.
- **Explosion resistant design**—in order for the equipment to withstand the max explosion pressure, or the reduced explosion pressure brought about by the use of explosion relief venting or explosion suppression.
- **Explosion suppression**—by preventing an explosion attaining its max explosion pressure by rapidly injecting extinguishing agents into the equipment in the event of an incipient explosion.
- **Explosion isolation** (Prevention of flame and explosion propagation)—by using special devices to prevent the propagation of explosions, e.g. through pipes, breathing devices or through filling and emptying lines.

This chapter provides a brief introduction to the abovementioned mitigating measures.

T. Jespen, *ATEX—Explosive Atmospheres*, Springer Series in Reliability Engineering, DOI 10.1007/978-3-319-31367-2_6

6.2 Explosion Relief Venting

Explosion venting can reduce the explosion overpressure by the opening of weak areas in the walls of the enclosure at an early stage of the explosion and letting some of the content escape to an area where its emergence can be accepted from a health and safety point of view. The overpressure inside the enclosure is thus lowered to a reduced pressure (normally noted as P_{red}) below the design strength of the enclosure, including all parts which can be exposed to the explosion pressure.

Note: The method only limits the explosion overpressure. Where flame and pressure effects—and possible toxic effects—outside the enclosure are to be expected, suitable precautions may be required.

The area of the vent is the most important factor that determines the value of P_{red} in a vented explosion (EN 14491 contains relevant design calculations for dust explosion venting, and NFPA68 may also be helpful).

In order to illustrate the design calculations and the parameter dependence, the following describes for dust explosions the vent area (A_V) calculation for an isolated enclosure (no explosion propagation from one enclosure to another through connecting pipelines) where the max reduced pressure ($P_{red,max}$) falls in the interval between 0.1 and 2.0 barg:

$$A_V = B \times (1 + C \times \log_{10}(L/D))$$

in m^2 where

$$B = \left(3.264 \times 10^{-5} \times P_{max} \times K_{St} \times P_{red,max}^{-0.569} + 0.27 \times (P_{stat} - 0.1) \times P_{red,max}^{-0.5}\right) \times V^{0.753}$$

and

$$C = -4.305 \times \log_{10}\left(P_{red,max}\right) + 0.758$$

where

L/D	Height to diameter of the enclosure
P_{max}	Max explosion pressure (bar)
$P_{red,max}$	Max reduced explosion pressure (bar)
P_{stat}	Static activation pressure (bar)
K_{st}	Dust explosion constant (bar × m × s^{-1})
V	Volume of the enclosure (m^3)

Note: When $P_{red,max}$ is above 1.5 barg, $A_V \approx B$.

The equation has the following limits if validity:

- Enclosed volume between 0.1 and 10,000 m^3.
- Static activation gauge pressure (P_{stat}) of the venting device between 0.1 and 1 bar.
- Max explosion gauge pressure (P_{max}) between 5 and 10 bar for a dust $K_{st} \leq 300$ bar m s^{-1}.
- Max explosion gauge pressure (P_{max}) between 5 and 12 bar for a dust K_{st} between 300 and 800 bar m s^{-1}.
- Homogeneous dust cloud.
- Atmospheric conditions.
- L/D ratio not greater than 20, or min vent area greater than the cross-sectional area of the vessel for end venting.

An example
A dust with a K_{st} of 175 (bar × m × s^{-1}) and a P_{max} of 9 (bar) handled in an enclosure of 128 m^3 with L/D of 6, and not to be exposed above explosion pressures of 0.5 bar.

The area (A_V) of the required explosion vent having a P_{stat} of 0.1 bar is calculated to be:

$$\begin{aligned} B &= \left(3.264 \times 10^{-5} \times 9 \times 175 \times 0.5^{-0.569} + 0.27 \times (0.1 - 0.1) \times 0.5^{-0.5}\right) \times 128^{0.753} \\ &= 0.0763 \times 38.6126 = 2.9461 \\ C &= -4.305 \times \log_{10}(0.5) + 0.7589 = 2.0548 \\ A_V &= 2.9461 \times (1 + 2.0548 \times \log_{10}(6)) = 7.66\,\mathrm{m}^2 \end{aligned}$$

With a K_{st} of 85, A_V is calculated to 3.72 m^2.

Vent closures may take the form of panels or doors, and they may be reusable following an explosion, or they may have to be replaced after an opening.

Where it is important to avoid oxygen entering the equipment to any great extent, the vent closure must be able to close again automatically.

Vent closures are considered as autonomous protective devices according to ATEX Equipment directive 2014/34 when separately placed on the market, and as such required to comply with the relevant essential health and safety requirements (EHSR) and to be tested.

6.2.1 Use of Vent Ducts

Vent ducts are used to direct a vented explosion to a safe area, but obstruct the venting process giving rise to an increased reduced explosion pressure.

The max reduced explosion overpressure with vent duct ($P_{duct,red,max}$) can be estimated with the following formula (see EN 14491 concerning validity range and supplementary information):

$$P_{duct,red,max} = P_{red,max} \times \left(1 + 17.3 \times \left(A \times V^{-0.753}\right)^{1.6} \times L\right)$$

where

$P_{duct,red,max}$ is the reduced pressure with vent duct (bar)
$P_{red,max}$ is the reduced pressure without a vent duct (bar)
A is the vent area (m^2)
V is the vessel volume (m^3)
L is the vent duct length (m).

6.2.2 *Effects External to the Vent*

The specification of a sufficient vent area is not the only aspect of explosion venting which also includes an appropriate handling of the hazards that may arise from the venting process.

These hazards include:

- Flame effects
- Pressure effects
- Material emissions
- Recoil forces

Flame length estimation

Based on dust explosion experiments, the max flame length ($L_{f,max}$) emitted from a vent can be estimated by:

$$L_{f,max} = C \times V^{1/3}$$

where

C is a constant taking account of the vent direction:

C = 8 for horizontal explosion venting, and
C = 10 for vertical explosion venting

V is the volume of the enclosure.

In practice, no flame length greater than 30 m has been measured, even for large volumes, which therefore should be regarded as the upper limit for any estimation.

Concerning the exact area of validity for the estimation, see EN 14491.

External pressure estimation

The max external overpressure due to pressure effects by a dust explosion inside the vented enclosure—and the explosion of the dust cloud outside the vent—can be estimated by the formula:

$$P_{ext,max} = 0.2 \times P_{red,max} \times A_V^{0.1} \times V^{0.18}$$

where

$P_{ext,max}$ is the max external overpressure (bar)
$P_{red,max}$ is the max reduced explosion overpressure (bar)
A_V is the vent area (m^2)
V is the volume of the enclosure (m^3)

Material emissions

Personnel must of course be excluded from the discharge areas of vents or vent ducts during normal operation. Deflectors can be installed to guide the vented explosion to a safe place.

Explosion venting inside buildings requires vent discharging into enclosed areas from which personnel are completely excluded when the plant is operating. Under no circumstances must secondary explosions be allowed to develop.

The basic requirement to vent into safe places may in some cases be nearly impossible to achieve without implementing very expensive plant modifications. In these cases the problem can be alleviated by the use of special techniques, as for example the use of deflector plates.

A deflector plate is simply a plate that is used to influence the path of the external flame and reduce the extent of the hazardous area external to the vent—having only negligible effect on the reduced explosion pressure.

The area of the plate should be at least three times the area of the vent, and be inclined at least 45°–60° to the horizontal, to deflect the ejected flame upwards.

The plate should be placed at a horizontal distance from the centre of the vent of 1.5 × D, where D is the diameter of the vent.

Recoil forces

Recoil forces are generated during explosion venting by ejection of material from the vent opening.

The max recoil force can be estimated using the following formula:

$$F_{recoil,max} = 119 \times A_V \times P_{red,max}$$

where

$F_{recoil,\,max}$ is the recoil force (kN)
A_V is area of the vent (m^2)
$P_{red,max}$ is the max reduced explosion overpressure (bar)

Installation of vents of equal area on opposite sides of a vessel can in some cases compensate for recoil forces, but imbalances due to a non-simultaneous opening of the vents has to be considered at the design stage.

The design of certain support structures for vented vessels is facilitated based on an estimation of the duration of the recoil forces:

$$t_R = 10^{-4} \times (K_{st} \times V)/\left(A_V \times P_{red,max}\right)$$

where

t_R is the duration time of the pulse (s)
K_{st} is the dust K_{st} value (bar × m × s^{-1})
V is the volume of the vessel (m^3)
$P_{red,max}$ is the reduced explosion pressure (bar)
A_V is the vent area (m^2)

The impulse transmitted by the recoil force is approximated by:

$$I_R = 0.52 \times F_{recoil,max} \times t_R = 62 \times A_V \times P_{red,max} \times t_R$$

where

I_R is the impulse (kN × s)
A_V is the vent area (m^2)
P_{red} is the reduced explosion pressure (bar)
t_R is the duration time of the impulse (s).

6.3 Explosion Resistant Design (Explosion Containment)

Using explosion resistant equipment which is capable of containing and withstanding the likely explosion pressure (be it the estimated max or a reduced explosion pressure) without rupture, is one way of mitigating the hazardous effects of an explosion.

Explosion resistant equipment is categorized either as:

- **Explosion pressure resistant** equipment, which is designed to withstand the likely explosion pressure without becoming permanently deformed (i.e. is reusable), or
- **Explosion pressure shock resistant** equipment, which is designed to withstand the likely explosion pressure without rupturing, but permanent deforming must be foreseen.

Explosion containment is often a suitable option for mills and other plant equipment of small volume which can be built to withstand the estimated explosion pressure without excessive cost.

The method is also used where a risk assessment points to the fact that emission of the process material or combustion products presents an unacceptable risk.

A reasonable confidence in the estimated explosion pressure is an evident prerequisite for implementing this mitigating measure.

Specifications for the design of pressure resistant and pressure shock resistant equipment can be found in EN 14460 which contains references to the EN 13445 series.

6.4 Explosion Suppression

It takes typically 30–100 ms after ignition before destructive explosion pressures emerge.

Explosion suppression uses this time delay by a very early detection of the incipient explosion followed by the discharge of sufficient suppressant into the developing explosion with the effect that the explosion is extinguishing before destructive pressures have developed.

Explosion suppression is generally used when explosion venting or containment is not a practical solution, and especially where emissions outside equipment would lead to toxic exposures. Sometimes explosion suppression is used in combination with relief venting.

Pressure detection devices are available, which located at the vessel boundaries can detect reliably the onset of an explosion within milliseconds. When combined with an appropriately designed control system to activate the injection of suppression material, and to shut down the relevant part of the plant in the event of a critical fault, a reliable explosion mitigating system can be in function for many situations.

Common suppressants are:

- Dry powders which are usually variants of the proprietary dry chemical fire extinguishants that have been conditioned to have a low median particle size and a high degree of fluidity. Powders include alkali metal bicarbonates, ammonium phosphates and substituted areas. Applicable for explosion dust class St 1, St 2 and some St 3.
- Fluorinated hydrocarbons such as FE36 and FM200. Applicable for explosion dust class St 1.
- Water, having a satisfactory suppressing effect for many dust explosions (Freezing may be a problem). Applicable for explosion dust class St 1 and St 2.

6.5 Explosion Isolation (Explosion Decoupling)

Explosion isolation is a technique that prevents an explosion pressure wave and flame (complete isolation), or only flame (partial isolation), from propagating via connecting pipes or ducts into other parts of the plant.

The systems can be either an active type, which requires a means of detecting the explosion and initiating an action to execute the isolation, or a passive type that requires no additional equipment in order to fulfil its function.

Some explosion isolation devices:

- Deflagration arresters (for gases) that prevent the transmission of explosion flames, withstand the explosion pressure and temperature stress of an deflagration.
- Flame arresters (for gases) that prevents the transmission of explosions by flame.
- Detonation arresters (for gases) that can withstand the mechanical and thermal stresses of detonations and prevent the transmission as well as act as deflagration arresters.
- Extinguishing barriers (for gases and dust) that prevent explosion propagation through pipes and ducts by the injection of extinguishing agents activated by appropriate detectors. The propagation of the pressure wave originating from the already burnt mixture up to the barrier, is not affected by this technique, and has to be taken into account.
- Rapid-action valves and flaps (for dust) that prevent flame and pressure propagation in pipes and ducts by closing in a sufficient short time—effected by means of an actuating mechanism initiated by detectors or by the explosion pressure wave itself.
- Rotary valves (for dust) that prevent flame and pressure propagation by stopping the rotor automatically by means of a detecting system in case of an explosion, in order to ensure that the discharge of the burning product is prevented.
- Explosion diverters (for dust) that prevent explosion propagation by changing the flow direction while—at the same time—relieving the explosion pressure, typically constructed as a duct where in the flow direction, a pipe bend is concentrically introduced into a segment of larger diameter containing a relief mechanism.
- Material chokes (for dust) where the actual product prevents the propagation of an explosion, e.g. by the use of level controls or removal of a section of the screw in a screw conveyor.

Chapter 7
ATEX—Risk Assessment

7.1 Introduction

As mentioned in Chap. 2, compliance with ATEX Workplace directive 1999/92 requires the employer (with or without support from special experts) to perform an explosion risk assessment, taking account at least of:

- The likelihood that explosive atmospheres will occur and their persistence,
- The likelihood that ignition sources will be present and become effective,
- The possible process conditions and possible interactions, and
- The scale of the anticipated effects (consequences) of an explosion.

The risk assessment is then to be followed by a risk evaluation, which must lead to hazard/risk minimizing activities if the risk is graded as unacceptable or non-compliance with the "principles of integrated explosion safety" is realized. In case of hazard/risk minimizing activities being implemented, the risk assessment procedure must be repeated.

7.2 Likelihood of an Explosive Atmosphere

The likelihood of an explosive atmosphere being present and its duration is what determines the zone classification of the area (space) in question—see Chap. 2.

The first step in a systematic explosion risk assessment is therefore to make an inventory of all flammable substances (used and produced) present at the plant. The available Safety Data Sheets (SDS/MSDS) have an important role to play in that respect.

The next step is to evaluate the degree of possible evaporation/dispersion of the flammable substances (gases, vapours, mists and dusts). The evaluation must include all relevant information about the process conditions and the explosion risk

T. Jespen, *ATEX—Explosive Atmospheres*, Springer Series in Reliability Engineering, DOI 10.1007/978-3-319-31367-2_7

parameters of the substances (see Chap. 3). All possible points of release must be identified.

Where only very small quantities of a flammable substance are handled, the atmosphere may be assessed to be non-hazardous, and no further actions are needed. Non-hazardous areas in relation to explosive atmospheres are generally identified as areas not requiring special precautions to protect the health and safety of the workers concerned.

In all other situations (where hazardous areas have been identified) it is recommended to consider the technical possibilities for preventing the occurrence of an explosive atmosphere, e.g. by inserting to a level <25 % of LEL (under all foreseeable conditions).

The remaining hazardous areas must then be classified in terms of zones based on an estimate of the duration and extent of the explosive atmosphere. The assessment should cover all operational conditions at the plant such as normal operation, including start-up and shut-down, maintenance, commissioning, decommissioning and reasonable malfunctions. It is very important that operations other than normal are assessed.

Note: Area classification (zoning) does not take account of ignition possibilities and the consequences of an explosion. Zoning is only about hazard identification as far as the likelihood of the presence of an explosive atmosphere is concerned.

The subsequent Chaps. 8– 10 deals with ATEX zone classification concerning oil fluids, natural gas and dusts.

7.3 Likelihood of an Effective Ignition Source

A possible ignition source may emanate from the equipment placed in ATEX zones, materials entering the process or be generated by the process, or come from external sources.

Ignition sources are classified according to the likelihood of their occurrence (see EN 1127-1):

- Sources that can occur continuously or frequently,
- Sources that can occur in rare situations, and
- Sources that can occur in very rare situations.

This classification is equivalent to the following equipment ignition source classification (used in ATEX equipment categorization):

- Sources that can occur in normal operation (which must be prevented for an ATEX equipment to fall in **category 3**),
- Sources that can occur solely as a result of malfunctions (which also must be prevented for an ATEX equipment to fall in **category 2**), and
- Sources that can occur solely as a result of rare malfunctions (which also must be prevented for ATEX equipment to fall in **category 1**).

Generally, you will meet the following definitions regarding ignition sources in the literature:

- **Possible ignition source** meaning any kind of ignition source:

  Hot surfaces, such as the walls of dryers, steam pipes, enclosures for electrical apparatus, which can ignite gases, vapours and dusts. Binding friction in moving parts, particularly at high speeds, can generate sufficient heat to cause the ignition of flammable atmospheres. In the case of powders, even slow speed may generate heat, which can cause powder to degrade and smoulder or burn and create a potential source of ignition. Poorly maintained bearings, agitators, and other moving equipment, are common sources of frictional heating.
  Flames and hot gases (including hot particles), including hot work such as welding, cutting and burning.
  Mechanically generated sparks produced by glancing impacts between materials can cause ignition. In practical terms, only sparks produced by the thermite reaction between the light metals or their alloys and rusty steel present a serious risk, unless very sensitive flammable materials are involved.
  Electrical apparatus (sparks). Arcs and sparks from electrical apparatus may ignite flammable concentrations of vapor or dust in air. Selection of electrical apparatus must always be suitable for the zone in question.
  Stray electric currents, cathodic corrosion protection.
  Static electricity (see subsequent part 6).
  Lightning.
  Radio frequency waves (waves from 10^4 to 3×10^{12} Hz). If the electromagnetic field is powerful enough and if the receiving aerial is sufficiently large ignition may occur.
  Electromagnetic waves (waves from 3×10^{12} to 3×10^{15} Hz). Sunlight can trigger an ignition if objects cause convergence of the radiation, and laser radiation may contain sufficient energy to cause ignition, also as an unfocussed beam.
  Ionizing radiation.
  Ultrasonics. In extreme cases, ignition may be induced by the heating caused by energy absorption from ultrasonic sound waves.
  Adiabatic compression and shock waves. Compression can create high temperatures, which can cause ignition of an explosive atmosphere (and deposed dust).
  Exothermic reactions, including self-ignition of dusts.

- **Equipment related ignition source** meaning any "possible ignition source" caused by the equipment (**own ignition source**) regardless of its ignition capability.
- **Potential ignition source** meaning an "equipment related ignition source" which may ignite an explosive atmosphere (The probability of becoming effective determines the equipment category).

- **Effective ignition source** meaning a "potential ignition source" able to ignite an explosive atmosphere taking the probability of occurrence into consideration (in normal operation, expected malfunction or rare malfunction).

Concerning ATEX equipment covered by the ATEX Equipment directive 2014/34, the selection of equipment—unless the overall risk assessment indicates otherwise—must comply with (underling that the higher the likelihood of an explosive atmosphere being present, the less likely the occurrence of an effective ignition source must be) the requirements stated in Table 7.1.

Whereas the assessment of the likelihood of ignition sources related to the equipment and the equipment categorization (see Chap. 12) fall under the equipment manufacturer's obligations, the identification and risk assessment of other relevant ignition sources rest primarily with the employer (plant owner).

The following types of equipment warrant special attention at the risk assessment:

- Fans
- Mixers
- Belt conveyors
- Mills
- Drag link conveyors
- Screw conveyors
- Rotary valves

The following equipment and plant data can be of importance at the risk assessment:

- Speed and power of moving parts
- Materials of construction
- Lubricated vs. seal bearings/gearboxes
- Type of seals, seal face materials
- Expected temperature rise
- Electrostatic properties and grounding systems
- Operating environment
- Mitigation systems, e.g. instrumented trips, metal detection, alignment sensors, explosion protection, etc.
- Drawings.

Table 7.1 Required equipment selection according to zone

Zone	Equipment
Zone 0 and zone 20	Category 1
Zone 1 and zone 21	Category 2 (and category 1)
Zone 2 and zone 22	Category 3 (and category 1 and 2)

The equipment must of course be suitable for gases, vapours or mists and/or dust, as appropriate—communicated by a "G", "D" or "DG"

7.4 Comprehensive Risk Assessment

The risk assessment must include appropriate considerations concerning all the present installations, substances and processes and their possible interactions.

Areas which are or can be connected to zone classified areas must be included in the risk assessment.

7.5 Anticipated Effects of an Explosion

The estimation of the possible effects of an explosion can be structured according to the effects of:

- Flames
- Thermal radiation
- Pressure waves
- Flying debris
- Release of hazardous materials
- The specific consequences caused by abovementioned effects will depend on:
- Chemical and physical properties of the flammable substances
- Quantity and confinement of the explosive atmosphere
- Geometry of the surroundings
- Strength of enclosure and supporting structures
- Protective equipment worn by the endangered personnel
- Physical properties of the endangered objects.

7.6 Some Aspects of Controlling Risks Due to Static Electricity

Note: Although safety at a specific project has been based primarily on other strategies, it is always necessary to control ignition sources as far as reasonably practicable.

The possible ignition source "static electricity" deserves special focus because the generation of static electricity cannot be prevented as its intrinsic origins are present at every interface. But the accumulation of static charges can be prevented under many circumstances by grounding or bonding, or e.g. by humidification.

Important sources of static electricity charge generation are:

- Fine particles or granular materials passing through chutes or via pneumatic conveyers,
- Steam, air, or gas flowing from any opening in a pipe or hose, where steam is wet or the air or gas stream contains particulate matter,
- Non-conductive power or conveyor belts in motion,
- Moving vehicles, and
- All motion involving changes in relative position of contacting surfaces, usually of dissimilar liquids or solids.

The aim of most static control measures is to provide a means whereby the separated charges can recombine harmlessly before sparking potentials are attained, or to avoid spark gaps where hazardous discharge could occur.

Electricity moves through conductors easily but can only flow with difficulty or not at all through or over the surface of non-conductors (insulators).

Non-conductors include gases, glass, rubber, amber, resin, sulphur, paraffin, and most dry petroleum oils and many plastic materials (e.g. nylon, PVC, and PTFE etc.). When electricity is present on the surface of a non-conductor, where it is trapped or prevented from escaping, it is called static electricity.

Electricity on a conducting body that is in contact only with non-conductors is also prevented from escaping and is therefore "static".

Note: Electricity cannot be created or destroyed, and the manifestation of either "–" or "+" charge indicates that somewhere else there exist the equal and opposite charge (although this may be the earth).

7.6.1 Charge Generation

The generation of charge can be divided into three different methods:

(1) Contact charging

Electrostatic charging occurs when solids or liquids move with respect to the materials with which they are in contact.

At liquid handling:

- Flow (especially two-phase liquid/liquid or solid) through pipes, valves etc. and especially fine filters
- Pouring from containers (e.g. buckets), particularly with low conductivity liquids
- Stirring/agitating two-phase liquids (e.g. water in hydrocarbons) or solid/liquid mixtures
- The flow of two-phase gas/liquid mixtures (e.g. wet steam, condensate in gas)
- Atomisation of conductive liquids (e.g. water washing, spray painting)
- Gravity settling of liquids or solids from liquid suspensions

At powder handling:

- Size reduction (e.g. milling, grinding)
- Some drying processes (e.g. fluid bed drying, spray drying)
- Pneumatic transfer
- Sieving
- Emptying and filling powder from/to bags or drums

In other situations:

- Rubbing insulating surfaces
- Walking on insolating floors
- Removing synthetic clothing
- Passing of insulating conveyor belts over rollers
- Unwrapping plastic film
- Working close to a charged insulator
- Pulling plastic liners out of drums

(2) Induction charging

This is only important when a conductive object is poorly earthed, and it arises because of the object's lose proximity to another charged object (e.g. an operator close to a water slug falling through a charged mist during tank washing). Personnel can become charged by induction, but if a person is wearing conductive footwear and standing on static-dissipative flooring, charge will dissipate to earth as soon as he is removed from the influence of the electric field. Only materials which are electrically conductive become charged by induction.

(3) Contact charging

Objects which are physically connected to other electrostatically charged objects can become charged simply by virtue of charge being passed on to them.

7.6.2 Electrostatic Charge Accumulation

A potential hazard is created when an electrostatic charge accumulation arises which will happen when the rate of charge generation exceeds the rate of charge decay from an object.

Typical charge accumulation scenarios are:

(1) Isolated metal parts

The ability of an isolated metal part to accumulate charge depends on its resistance to earth. In general terms, this resistance has to be above 10^6–10^8 ohms (depending on rate of charge generation) in order to create hazardous potentials.

Spark energies in the region of 50–500 mJ are not uncommon from isolated metal parts.

(2) Personnel

An ignition risk due to electrostatically charged personnel may exist when flammable atmospheres with minimum ignition energies less than 100 mJ are present. Personnel may become charged if they wear insulating shoes or the floor is insulating (resistance to earth $>10^8$ ohms).

Sparks from isolated conductors such as metal parts and personnel are among the most incentive types of discharge.

(3) Non-conducting materials

Non-conductive materials (e.g. rubber, PVC, polyethene, PTFE, glass and fibre laminates) are defined as materials having a volume resistivity $>10^9$ (ohm × m) or a surface resistivity $>10^9$ (ohm)—measured at ambient temperature and 50 % relative humidity, and these materials may be capable of accumulating electrostatic charge.

Materials such as wood, paper are not usually insulating and therefore do not normally accumulate charge.

Electrostatic discharge from insulating solid materials can be incentive to most flammable vapour/gas atmospheres and (probably) to some sensitive dust clouds.

(4) Bulk liquids

Charge generated during flow is entrained when the liquid enters a storage vessel. Turbulent flow generates more charge than laminar flow. The charge per unit time carried by the liquid as it enters a receiving vessel decreases as the liquid conductivity increases. Retaining a hazardous charge is not possible with conductivities >50 pS/m.

Charge is easily accumulated when the conductivity is of the order of a few pS/m or less (relaxation time ≥20 s), but it is usually dissipated to rapidly to produce hazardous potentials when the conductivity is >50 pS/m (relaxation time ≤0.4 s).

The relaxation time is a measure of the time required to dissipate charge.

Low conductive liquids (conductivity ≤50 pS/m) are for example: Toluene, xylene, hexane, heptanes, carbon disulphide and petroleum.

(5) Bulk powder

The ability of charge to accumulate on powders is dependent upon the Powder Volume Resistivity (ohm × m) of the powder, and the charge relaxation time, which is a function of the Powder Volume Resistivity.

Powders are divided into three categories of resistivity:

- Low resistivity: $<10^6$ (ohm × m)
- Medium resistivity: 10^6–10^{10} (ohm × m)
- High resistivity: $>10^{10}$ (ohm × m)

Low resistivity powders are rare because metal powders will eventually oxidize.

When Powder Volume Resistivity are $>10^9$ (ohm × m), there is no electrostatic risk in earthed metal plants provided that the Minimum Ignition Energy is >10 mJ.

Charged powder in a plastic container can give rise to very energetic electrostatic discharges.

(6) Liquid mists and powder clouds

Production of a liquid mist usually involves an atomization process which in turn results in a charged liquid.

Liquid mists can also be produced from an aerosol, cleaning operation etc.

A charged mist produces high potentials and the max, in the case of a tank, would be in the centre.

Charged liquid mists and dust clouds can ignite sensitive gases, vapours, mists and dust clouds, but no hazard if the Minimum Ignition Energy exceeds 25 mJ.

(7) Gases

Pure gases do not charge, but significant charges can be generated if the gas contains suspended particles or droplets.

7.6.3 *Discharge Mechanism*

Six types of electrostatic discharge can be identified (see Table 7.2) of which five can occur inside plant and processes. The sixth type—a lightning type of discharge—has not been established as relevant at the plant level.

Table 7.2 Discharge types

Discharge type	Origin	Energy	Incentivity
Spark	Isolated conductor	Hundreds of mJ for unearthed metal. Up to 50 mJ from isolated people	Vapours, gases, dusts
Brush	Insulating plastics	Up to 4 mJ	Vapours, gases, perhaps some sensitive dusts
Propagating brush	Insulator in closed contact with earthed metal	Up to several J	Vapours, gases, dusts
Cone	Bulked powder—cone of material	Fine powder: about 10 mJ Granules, pellets: Much higher	Vapours, gases, dusts
Corona	From sharp edges and corners	$\ll$0.1 mJ	Vapours
Lightning	Atmospheric conditions	Many hundreds of J	

Table 7.3 Examples of possible spark energies

Charged object	Capacitance (pF[a])	Stored energy (mJ[b])
Small metal items (scoop, hose nozzle)	10–20	1–2
Small containers (bucket, 50 l drum)	10–100	2–10
Medium containers (250–500 l)	50–300	5–30
Major plant items (reaction vessels, etc.)	100–1000	10–100
Person	100–300	10–30
Road tanker	Up to 1000	Up to 100

[a]1 pF = 1×10^{-12} F
[b]Voltage of 20 kV (20×10^3 V), voltage maybe higher or lower

The discharge energy varies according to the type of discharge mechanism, and the energy discharged can range from ≪0.1 mJ to several J (Table 7.3).

Hazards—static shocks:

1/10 mJ—Undetectable by touch—ignites H_2/CS_2 in air vapours in oxygen
1/4 mJ—Still undetectable—ignite vapours and gases in air
1 mJ—Detectable—ignite sensitive dust clouds
10 mJ—Prickling sensation—ignite dust clouds
100 mJ—Muscular contraction—very incendive
1 J—Sharp contraction—potent incendivity.

7.6.4 *Control Methods*

7.6.4.1 Metallic and Conductive Components

All fixed items should be reliably earthed with a resistance to earth <10 ohms.

In special cases a resistance to earth of 10^6 ohms can be acceptable.

All mobile equipment should be earthed by earthing leads.

Any Flexible IBCs (big bags) used should be of type C and earthed at the point of use.

The reliability of the earthing should be checked at suitable intervals, e.g. annually.

7.6.4.2 Flexible Hoses and Connections

Flexible connections and hoses can cause two problems in flammable atmospheres. They can cause metal items to be isolated from earth and the wire spiral in some flexible hoses can be a source of charge accumulation and subsequent spark.

Conductive hoses have a resistance of <10^3 ohm/m which prevent electrostatic discharges, but stray currents can be conducted which could create an ignition hazard if the stray current is interrupted. There should be no isolated metal items, such as end couplings, reinforcement helixes and/or outer sheathing.

Semi-conductive hoses have a resistance (10^3–10^6 ohm/m) which is low enough to dissipate electronic charges but is high enough to restrict stray currents.

Generally, semi-conductive hoses are preferred to conductive as they protect against electrostatic charge generation on the hose, as well as against inductive sparks due to interruption of stray currents.

Non-conductive hoses are not capable of dissipating electronic discharges. Electrostatic discharges on the hose or isolated conductive components may therefore lead to incendive discharges. They are not recommended for flammable atmospheres.

7.6.4.3 Personnel and Flooring

The body is an electrical conductor which when isolated from earth can store charge and may be a source of ignition. An unearthed operator can be a source of discharge up to 50 mJ (theoretically 100 mJ) having become charged by own movements or by being in contact with or near to other charged material.

Work where flammable atmospheres may occur requires persons to be earthed by means of suitable footwear and floors.

Conductive footwear (resistance less than 5×10^4 ohm) can be worn if there is no risk of electrical main shock. Antistatic footwear with a sole resistance in the range 5×10^4–10^8 ohm avoids the risk whilst ensuring adequate earthing.

Floors should have a resistance to earth less than 10^8 ohm from any point of them (Note: Contamination can affect the resistance significantly).

For common solvents with an MIE (Minimum Ignition Energy) ≈0.2 mJ, it is not essential to wear antistatic clothing, but it is essential that the clothing is not removed in the presence of flammable vapour atmospheres and that the clothing is close fitting and fastened down the front.

7.6.4.4 Liquids

Pipeline velocities should not exceed 1 m/s for highly resistive solvents, i.e. with conductivity <50 pS/m, including all multiphase systems.

It is recommended to keep pipeline velocities below 7 m/s for all other solvents (as single-phase systems).

As both spraying and splash filling can generate significant amounts of charge, these methods should be avoided by the use of e.g. dip-pipes.

The spraying of electrically conducting liquids can result in significant electrostatic charging, and the formation of mists. In large vessels (>30 m^3 approx.), the formation of mist by the spraying of any liquid may result in a risk of ignition.

Restricting the flow velocity to <2 m/s will normally prevent spraying and mist formation. Where multiphase systems are involved, the flow velocity should not exceed 1 m/s.

7.6.4.5 Powders

Control of the rate of charge generation in powders is difficult. The level of charging depends more on the work done on the material than its chemical nature. Choosing handling and processing methods that involve as little work on the powders as possible will reduce the charging.

It is considered that a hazard could only exist with powdered products of MIE <10 mJ provided the powder is in contact with earth. However, for earthed pneumatic transfer pipe of diameter <1 m, there is no hazard of ignition from electrostatic discharges originating from the dust cloud within the pipe for materials with MIE >0.2 mJ.

7.6.4.6 Non-conductive Components

Non-conductive equipment (e.g. plastic) cannot be earthed. This leads to the retention of electrostatic charge on the equipment and products contained in it, for example generated by rubbing friction.

The energy of the discharge that can be drawn from these surfaces is limited (normally less than 4 mJ and always less than 10 mJ), but this is sufficient to ignite most gases and vapors and some very sensitive dust atmospheres.

In addition, plastic components can cause conducting plant items to be isolated from earth so that the item can accumulate charge and become an ignition risk with potentially greater energy.

It is recommended to avoid the use of non-conducting equipment and plant item in areas where flammable gases, vapors and sensitive dust atmospheres with MIE below 5 mJ are or may be present. With vapors, insulating plastics should never be used in Zone 0 or Zone 1.

Furthermore, in the case of dusts, insulating plastics should not be used where dust clouds with MIE less than about 10 mJ may be formed, or in high charging rate systems such as pneumatic transfer.

7.6.4.7 Vessels

Charge can accumulate on bulk materials in vessels. In order to restrict this, vessel contents should be maintained in contact with earth. Where conducting unlined vessels are used, no steps other than earthing the vessel are needed. In vessels with insulating linings, it is necessary to install an appropriate earth contact inside the vessel as well as earthing the shell. In glass-lined vessels, an earthing focus is generally provided by the inclusion of an earthed tantalum plug situated at the base of the vessel or a tantalum tip to a temperature pocket.

These precautions are almost always confined to liquid systems, although similar procedures can be used with powders in particular circumstances.

7.7 Battery Charging Areas

Hydrogen will be released from batteries during charging. However release rates are extremely low and hydrogen is a very light gas that rapidly rises away from the charging area, and accumulation at low levels will not occur provided a reasonable level of general workplace ventilation is present. Accumulation at high-level roof peaks etc. may be of concern is adequate high-level ventilation is not provided.

Chapter 8
Zone Classification—Oil Fluids

8.1 Introduction

This chapter is primarily concerning hazardous areas around equipment handling or storing oil fluids belonging to Class II (1), Class III (1), Unclassified handled below T_F or category C—see definitions below.

The basic principles presented here are particularly based on the publications mentioned in "References".

As a general guide (see IP 15), zone classification is not needed when the max amount of flammable material that could be released, is below the quantities stated in Table 8.1.

But no hard and fast rule subsists as some materials are more potent than others.

Liquid oil fluids cannot burn. Combustion (the oxidation to H_2O and CO_2) of the fuel requires a sufficiently rich oil vapour-air mixture, which may be formed when the oil fluid is heated to the Flash point (T_F) **or** by the creation of a mist of fine oil droplets in the air.

If the fuel vapour and air is ignited and mixed during the combustion process, the reaction takes the form of a fire.

If the fuel vapour and air is ignited following a sufficient premixing, the oxidation reaction may proceed much faster, and an explosion may be the outcome.

Releases of oil fluids below their T_F may under certain conditions give rise to the formation of a flammable mist cloud, which constitute an explosion hazard that may be ignited by sparks of similar energy as for vapour ignition.

Flammable mists may also be created inside equipment due to splashing or agitation as part of the process conditions, and this mixture may in some cases be vented to the atmosphere.

Furthermore, mists and droplets that are not ignitable could eventually land on a hot surface, relative to the ignition temperature of the vapour, thus causing a fire/explosion hazard.

T. Jespen, *ATEX—Explosive Atmospheres*, Springer Series in Reliability Engineering, DOI 10.1007/978-3-319-31367-2_8

Table 8.1 Quantities not requiring zone classification

	Gas: equivalent volume at 1 Bar (l)	LPG (l)	Liquid at temperature > T_F (l)
Inside	50	5	25
Outside	1000	100	200

8.2 Classification of Oil Fluids

The key property of a flammable liquid fluid concerning fire and explosive risks is its volatility, expressed through its Flash point (T_F), since it will determine the extent of rapid vapour formation from any release.

Petroleum and flammable fluids are generally classified according to the Flash point (T_F)—except for LPG—with a further subdivision according to whether the liquids are handled above or below their T_F, as stated in Table 8.2.

It is recommended to apply a subdivision of the Unclassified as for Class III (and Class II) and regard the handling of Unclassified fluids above T_F as Class III (2) fluids.

Above-mentioned classification is often inadequate for a sufficient zone classification under typical processing conditions. This has lead to the introduction of the concept of “fluid category”, which is a crucial factor in calculation of the zone hazard radii (see Table 8.3).

Table 8.2 Oil fluid classes

Class	Fluids
Class 0	LPG (liquefied petroleum gases)
Class I	Having T_F < 21 °C
Class II (1)	Having 21 °C ≤ T_F ≤ 55 °C, handled below T_F
Class II (2)	Having 21 °C ≤ T_F ≤ 55 °C, handled at or above T_F
Class III (1)	Having 55 °C < T_F ≤ 100 °C, handled below T_F
Class III (2)	Having 55 °C < T_F ≤ 100 °C, handled at or above T_F
Unclassified	Having T_F > 100 °C

T_F: Flash point temperature

Table 8.3 Oil fluid categories

Category	Description
A	A flammable liquid that, on release, would vaporize rapidly and substantially
B	Any flammable liquid, not belonging to category A, but handled at a temperature sufficient for boiling to occur on release
C	Any flammable liquid, not belonging to category A or B, but which can, on release, be at a temperature above its flash point, or **form a flammable mist or spray**
G(i)	A typical methane-rich natural gas
G(ii)	Refinery hydrogen

Table 8.4 Some examples of oil fluids

Class	Examples	Typical T_F (°C)	Typical T_B (°C)
0	Ethylene, propylene	N/A	<0
I	Gasoline	–45	20–205
	Stabilised crude oil	<21	−1 to 380+
	Naphtha	–2 to 10	30–177
	Methanol	12	65
	Ethanol	14	78
II	Iso-butanol	28	108
	1-Butanol	37	117
	Turbofuel	>38	150–240
	Kerosine, regular	38–72	150–280
III	Gas oil	55+	250–360
	Automotive diesel fuel	55+	180–360
Unclassified	Atmosphere residues	>100	>350
	Heavy fuel oils	>100	>350
	Biodiesel	>130	
	Hydraulic oils	Generally > 180	
	Heat transfer oils	Generally > 200	

Note: Class II (1), Class III (1) and Unclassified handled below T_F do not fall into any category unless handled at conditions possible for the creation of a flammable mist, in which case the fluid belongs to category C.

Comments: Oil fluids as Heavy fuel oil, Diesel oil and Heat transfer oil, generally belong to the group consisting of Class II (1), Class III (1) and Unclassified handled below T_F (in this chapter designated "**standard oil fluids**"), when received for processing at plant equipment.

Note: Only oil fluids classified accordingly or belonging to category C can make direct use of the generic oil fluid zone classification resented in this chapter.

In other words, where a categorisation of the fluid during processing is appropriate, it is a prerequisite for the fluid to fall in category C (and no other category). Should the fluid be processed at temperatures above T_F, it may still belong to category C as long as the handling temperature stay below the Boiling point (T_B) (Table 8.4).

8.3 Zone Classification—The Simple Cases

As long as we are dealing with above-mentioned oil fluids (which in this chapter will be named "**standard oil fluids**") in Class II (1), Class III (1) and Unclassified stored or handled below T_F, at system pressures below 5 Bar (which cannot at release form a flammable mist or spray), areas around equipment are generally not regarded as hazardous areas requiring zone classification.

As an exception, recommended by IP 15, the area around vents on storage tanks should be Zone 1 classified with a radius of 1.5 m and the ullage space of the tankage to be Zone 0 classified.

This recommendation also takes care of the fact that T_F is not a reliable indicator of the presence of flammable atmospheres in the ullage space when for example Unclassified fluids are stored under confined heated conditions.

8.4 Zone Classification Key Terms

First of all we have the **Zone definitions** (as stated in ATEX directive 1999/92) repeated in Table 8.5, here including the dust classifications (20, 21 and 22) although without significance for oil fluids.

The (universal) guideline presented in Table 8.6, can be used at the zone classification task concerning the duration (t) of the occurrence of an explosive atmosphere.

Note: Zone classification only takes into consideration the type of explosive mixture and the likelihood of its occurrence (determined by the frequency and

Table 8.5 Zone classification definitions

Zone	Definition
Zone 0	A place in which an explosive atmosphere consisting of a mixture with air of flammable substances in the form of gas, vapor or mist is present continuously or for long periods or frequently
Zone 1	A place in which an explosive atmosphere consisting of a mixture with air of flammable substances in the form of gas, vapor or mist is likely to occur in normal operation occasionally
Zone 2	A place in which an explosive atmosphere consisting of a mixture with air of flammable substances in the form of gas, vapor or mist is not likely to occur in normal operation but, if it does occur, will persist for a short period only
Zone 20	A place in which an explosive atmosphere in the form of a cloud of combustible dust in air presents continuously, or for long periods or frequently
Zone 21	A place in which an explosive atmosphere in the form of a cloud of combustible dust in air is likely to occur in normal operation occasionally
Zone 22	A place in which an explosive atmosphere in the form of a cloud of combustible dust in air is not likely to occur in normal operations but if it does occur, will persist for a short period only

Table 8.6 Zone classification and duration of explosive atmosphere

Zone	Guideline
Zone 0 and 20	$t > 1000$ h/year or $t > 1$ h/shift
Zone 1 and 21	$10 < t < 1000$ h/year
Zone 2 and 22	$1 < t < 10$ h/year and $t < 1$ min/shift

Table 8.7 Grade of release

Grade	Definition	Guideline (hours/year)
Continuous grade of release	A release that is continuous or nearly so, or that occurs frequently and for short periods	$t > 1000$
Primary grade of release	A release that is likely to occur periodically or occasionally in normal operation, i.e. an anticipated release	$10 < t < 1000$
Secondary grade of release	A release that is unlikely in normal operation and, in any event, only infrequently and for short periods	$1 < t < 10$

duration of its occurrence, which outside equipment is influenced by the degree of ventilation of the area).

Regarding releases of flammable gas, vapor or liquid from a point source into the atmosphere (for example through vents and failure leaks), and closely connected to the zone classification conception, one must be familiar with the concept of **Grade of release** (which is not influenced by the degree of ventilation), see Table 8.7.

In general, a continuous, primary and secondary grade of release under unrestricted "open air" conditions will give rise to a Zone 0, 1 and 2 respectively.

Determination of the **zone extension** outside equipment, for example caused by failure leaks, is normally the trickiest aspect of the zone classification task, being influenced by the following factors:

- Fluid characteristics, especially vaporizing potential (flash point, boiling temperature and handling temperature), Lower Explosive Limit (LEL) and density
- Degree of ventilation of the area
- Rate and volume of the release, influenced by system pressure and release area (leak hole size).

A basic (calculated) dimension concerning a point source pressurised release (characteristic for a secondary grade of release) is the **hazard radius** (R_1), defined as the horizontal extent in open air of the hazardous area independent of ground effects. That means R_1 is determined as the calculated distance at which the concentration of flammable vapor has fallen to LEL, due to the dilution by the surrounding atmospheric air.

When the source of release is situated at a height (H) where $H > R_1 + 1$ m, ground effects can be neglected, and the three-dimensional hazardous area (sphere) is framed by a sphere around the source with a radius of R_1.

Where $H < R_1 + 1$ m, the "ground effect radius" (R_2) is also relevant and determines the extension of the hazardous area up to a height of 1 m above ground level.

The degree of **ventilation** of an area is a key factor in determining the correct zone classification outside equipment.

Two boundary cases (open area and enclosed area) and an intermediate case (sheltered or obstructed area) can be identified and are defined below.

An **open area** is defined as an outdoor area without stagnant regions where vapor is rapidly dispersed by wind and natural convection, i.e. by natural ventilation. Typically, air velocities at ground level will rarely be less than 0.5 m/s and will frequently be above 2 m/s.

An **enclosed area** is defined as any building, room or enclosed space e.g. cabinet, within which, in the absence of artificial ventilation, the air movement will be limited and any flammable atmosphere will not be dispersed naturally.

A **sheltered or obstructed area** is defined as an area within or adjoining an open area where owing to obstructions, natural ventilation may be less than in a true open area. Below-grade areas such as pits and trenches, areas within tank bunds and congested plant structures may fall into this category. Where the obstruction is such that natural ventilation is severely restricted, the area should be regarded as an enclosed area.

As a general recommendation it may be put forward that an area with restricted natural ventilation should be assigned a zone classification of one step higher than a similar area located in open air.

An **adequate ventilation** of an enclosed area is generally defined as a ventilation rate of 12 air changes/hour with no stagnant regions, which is considered as sufficient to ensure that a building containing secondary grade release sources can be properly classified as Zone 2 (as in the case of an open air situation). In large buildings some parts may even be regarded as non-hazardous.

Note: EN 60079-10-1 in its Annex B handles the ventilation aspect slightly different, and IGM/SR/25 on natural gas defines adequate ventilation, as ventilation sufficient to ensure that a gas concentration above 25 % of LFL is unlikely outside the classified zone(s)—see also Chap. 9.

Generally speaking, everywhere (outside equipment) a potentially explosive atmosphere may occur, a building should be designed to allow an appropriate draught (openings in the walls, etc.) or be equipped with good (adequate) artificial ventilation.

8.5 Zone Classification Technique

The basic aim of the zone classification approach is to minimize (to an acceptable level) the probability of the coexistence of an explosive atmosphere and an ignition source.

Therefore, where the probability of the occurrence of an explosive atmosphere is high, as in Zone 0, one must choose equipment having a very low probability of triggering an equipment related ignition source (also called an "own ignition source"), as equipment in ATEX Category 1 in accordance with ATEX directive 2014/34—see also Chap. 12.

In order to zone classify potential oil fluid explosive atmospheres correctly it is important to know the Flash point (T_F) of the fluid—or at least to which class the fluid belongs.

The present short guidelines only focus on "**standard oil fluids**", here defined as oil fluid belonging to Class II (1), Class III (1) or Unclassified stored or handled below T_F when received for processing. And where process conditions make a categorisation of the oil fluid appropriate (due to system temperature and/or pressure), the fluid is to be covered by category C.

In other words, only fluids which in combination with the process conditions will place the fluid in category C (and not other categories) are covered by these guidelines.

Regarding **hazardous areas outside equipment in open air**.

The maximum ambient temperature must be known. These guidelines assume in general a max ambient temperature of about 30 °C in line with IP 15, but concerning the "standard oil fluids" this temperature may be fixed to around 45 °C—see References.

Estimation of hazardous areas around vents on storage tanks may require information about vent diameter and vent rate (filling rate).

Where a pool of liquid may be generated due to spillage, the estimated equivalent pool diameter (L) must be stated (which together with the actual temperature conditions determine the relevant zone classification).

Point source release other than vents includes typically failure leaks from pump seals, valves, flanges, sampling and drainage points, instrument connections, releases from rotating machinery such as pumps and compressors. Zone classification will depend on the rate/volume of release (and the specific ventilation conditions) which is determined by fluid characteristics, system temperature and pressure and aperture size (leakage area).

Regarding **indoor hazardous areas outside equipment**:

Besides the above-mentioned, the ventilation of the room is the key aspect in zone classification.

In **subsequent** Sect. 8.7 zone classification basics related to the "standard oil fluids" will be mentioned for the following situations:

- Road tanker unloading in open air
- Storage tanks and vents
- Liquid pools surrounding storage tanks
- Point source releases (failure leaks) in open air, generating mist and/or pool, and
- Some aspects of indoor releases.

8.6 Zone Classification and Risk Assessment

Zone classification is about assessing/estimating the likelihood that explosive atmospheres will occur, their type and persistence and the extension of the hazardous area.

Zone classification is a key element of the explosion risk assessment obligation laid down on the plant owner (the employer) forming part of his "Explosion protection document", which must be drawn up prior to the commencement of work. The risk assessment obligation also includes an estimate of the likelihood that ignition sources will be present possessing sufficient energy to ignite the explosive atmosphere, and the scale of the anticipated associated consequences.

As mentioned in EN 60079-10-1 (Clause 4.2), a specific zone classification is to be supplemented by a risk assessment in order to decide about the basic equipment selection rules (combination of zone type and equipment category) ought to be deviated.

Zone classification is not an exact science, and in some situations it will be appropriate to operate on a higher "safety" level (choosing a stricter zone classification and/or the use of more safe equipment) or to use less safe equipment than specified by the legislation—based on sound risk assessment considerations.

8.7 Some Zone Classification Situations

This part only deals with oil fluids, here defined as "standard oil fluids".

8.7.1 Road Tanker Unloading (Open Air)

Regarding unloading, the tank area and where the creation of a liquid pool is possible, is classified up to a height of 1 m as a Zone 2, and pits and tunnels as Zone 1 areas (in conformity with the "one step up" rule mentioned in Sect. 8.4 under sheltered or obstructed area).

Connections from road tanker to unloading tank are classified as Zone 1.

8.7.2 Storage Tanks and Vents

Generally, the "standard oil fluids" are stored at temperatures below T_F and require no zone classification outside the tanks **but** in accordance (and based on prudence) with IP 15, the ullage space of the tankage is classified as Zone 0 with a 1.5 m Zone 1 around roof vents and other openings. This classification reflects the hazards connected with the tank filling situation (covering filling rates not exceeding 250 m^3/h). Under normal non-heated operation a Zone 2 classification around vents should be appropriate.

The Zone 0 and 1 classification (recommendation) takes into account the possibility of mist, spray formation of Class II (1) and III (1) fluids and Unclassified fluids in the storage tanks. Furthermore, the fact is that T_F is not a reliable indicator of the presence of a flammable atmosphere in the ullage space when these fluids are

Table 8.8 Hazard radius (R_1 in meters) from tank vent for fluid category C

Vent/filling rate (Nm^3/h)	Vent diameter (mm)			
	50	80	100	250
1000	3	4	4	6

R_1: The extension of a Zone 1—as ground effects usually can be neglected

stored under **confined heated conditions**. I.e. the stated zone classification also covers heated storage.

Where feasible, the area around vents is extended as a Zone 2 down to ground level.

Where a standard oil fluid falls into category C, Table 8.8 illustrates the Zone 1 extensions (radius) for a filling rate of 1000 Nm^3/h depending on vent diameter.

The area around the vent may be classified as a Zone 2 during normal operation where no tank filling takes place.

Storage tanks **holding category C fluids** as Class II (2), Class III (2) or Unclassified above T_F (including Class I fluids) should be assigned a Zone 1 area with a height of 3 m on the top. The zone should be extended 2 m down the shell from the roof edge (horizontally with a 3 m width), from where the area continues down as a Zone 2.

Inside tank area walls, a Zone 2 classification at a height of 1 m is recommended for tanks holding category C fluids, and for fluids outside category when the ambient temperature may exceed 50 °C. For category C fluids the height is extended to 3 m when temperatures above 50 °C are expected.

8.7.3 Liquid Pools Due to Spillage in Open Air

Where a liquid pool (equivalent diameter L) of a Class II (1), Class III (1) or Unclassified stored below T_F may be created, e.g. around a storage tank in open air, no zone classification is normally required for the area connected with the pool.

But it is good practice to restrict electrical apparatus within bunds to the minimum necessary, and pumps and their drives should not be situated within tank bunds.

For category C fluids (and fluids outside categorization at ambient temperatures above 50 °C) a Zone 2 with a height of 1 m is recommended. For category C fluids at ambient temperatures above 50 °C, the height of the zone should be extended to 3 m.

The area of the pool, expressed through its equivalent diameter (L) must be estimated based on the possible amount of spillage which is influenced by system pressure, leak area and duration together with the possibility of the spillage to spread in the surroundings.

Where the fluid belongs to category C, the recommended hazard radius (R) of the Zone 2 classification is determined by Table 8.9.

Table 8.9 Pool diameter and estimated hazard radius

Pool diameter L (m)	R (m)
$L < 5$	3
$5 \leq L < 10$	7.5
$L \geq 10$	15

L: Equivalent diameter, i.e. the diameter of the circle of equivalent area to the rectangle

Note: The hazardous radius (R) is to be regarded as an extension of the radius from the edge of the pool in the same direction, implying a radius of $\frac{1}{2}L + R$ of the Zone 2 (Pits and depressions to be Zone 1 classified).

8.7.4 Point Source Releases (Failure Leaks) in Open Air

Point source releases typically includes possible failure leaks from valves, flanges, instrument connections, releases from rotating machinery such as pumps and compressors and any areas where spillage from these sources could collect.

In properly designed, operated and maintained facilities, such point sources will generally result in secondary grade releases, not to be classified stronger than as a Zone 2.

Possible leak source releases of the "standard oil fluids" outside category C, i.e. Class II (1), Class III (1) and Unclassified handled below T_F and at system pressures below 5 Bar, are not considered as zone classification candidates, because the possible mist formation is regarded as negligible. But the presence of nearby hot surfaces where the oil release may settle down should be included in the risk assessment.

Point source release of oil fluids handled below their T_F (or above) may be in the form of a flammable mist or spray depending on the system pressure and hole size. The flammability of the mist depends in a complicated way on the ease of ignition of the fluid, the concentration (droplets plus vapour), volatility and the droplet size distribution within the cloud.

The size of the droplets depends on the release pressure, the properties of the liquid and the size and shape of the release opening. Normally, higher pressures and smaller openings will contribute to the generation of fine droplets (aerosols) and enhance the explosion risk.

Not all of the liquid released from a pressurised source might remain airborne. Thus, loss of fluid to the ground through "rain-out" may mitigate the hazard associated with two-phase releases.

According to the available information (based on release model calculations), rain-out dominates definitely at pressures below 5 Bar, leading to the conclusion that Point source releases of the fluids at system pressures below 5 Bar do not lead to a zone classification surrounding the source, but a pool may be generated.

As mentioned above, no zone classification is normally associated with this pool as long as the temperature is below T_F (see Sect. 8.7.3).

Although rain-out fraction has been calculated to 44 % of the mass released at 10 Bar, a pressure above 5 Bar has been fixed for categorizing the fluid as category C, thus being a point source release candidate surrounded by hazard radii.

Note: The 5 Bar borderline between mist and no mist formation should be regarded as a rather conservative value, and may be looked upon with some flexibility regarding more viscous liquids, especially when handled below the flash point.

The factors determining the extent of a hazardous area from a point source include vaporising potential of the fluid release, the degree of ventilation and the rate or volume of the release.

Besides the system pressure, knowledge about (or a sound estimation) of the possible hole size for determining an appropriate leak rate is of key importance for estimating the zone extension (See also IP 15, Annex C—Part 2 and 3).

A hole size of diameter 2 mm regarding pumps and of 1 mm for other equipment is often considered appropriate as default values based on equipment quality considerations and maintenance according to instruction manuals.

The likelihood of releases from flanges and valves is very low, so it may not warrant zone classification associated with an individual item, particularly if it is not operated at high pressures or temperatures. As a guide, only in cases with more than 10 possible leak sources within close proximity, the whole area should be Zone 2 classified (Opposite the general Zone 2 classification around a pump).

As mentioned above, a point source release may create a pool due to some rain-out, which in cases where the temperature is above T_F may create—depending on the specific surroundings—a need for a zone classification associated with the pool.

In "open air", the pool question may generally be neglected (ground effects to be handled through the R_2 radius), but where the pressure is only slightly above 5 Bar, the combination of the specific surroundings and the possible release spray may give rise to a more detailed assessment.

8.7.5 Some Aspects of Indoor Releases

Gas or vapor released to the atmosphere will eventually be diluted by dispersion in free air until its concentration is at a safe limit below the LEL (Lower explosive limit).

Where the release is not into completely free air, the specific air flow, or ventilation, is a very important factor to be taken into consideration.

Ventilation of a confined space is typically quantified by a single parameter—the number of air volume changes per hour.

As a general rule, all continuous and primary grade sources of release should never be located within sheltered or obstructed areas or enclosed areas (not acceptable practice). But in some cases an open area cannot fulfil the "Open air" requirements specified in Sect. 8.4 which then requires a more detailed evaluation of the specific circumstances.

An enclosed area with a continuous internal release source must be classified as a Zone 0.

An enclosed area with a primary internal release source must be classified as Zone 0 for inadequate ventilation and Zone 1 for adequate ventilation.

Secondary grade release sources may be present in enclosed areas, generally leading to a Zone 1 classification for inadequate ventilation and a Zone 2 classification for adequate ventilation.

The following **comments concern only secondary grade releases**.

As a basic guideline concerning adequately ventilated areas, it is recommended to classify the whole enclosed area as Zone 2 where a point source release (see Sect. 8.7.4) in the area of a category C fluid is regarded as a possibility due to failure leaks.

In large buildings (with large dimensions compared to calculated hazard radii) with adequate ventilation, it may be appropriate to classify some parts as non-hazardous, while other parts are Zone 2. In those cases it may be appropriate to enlarge the hazard radii to twice the calculated "open air" values.

Smaller areas of a generally adequately ventilated enclosed area, where the ventilation may be assumed to be at a lower level are candidates for a one step up zone classification (Zone 2 as the general classification of the room and Zone 1 in the small areas).

Handling of the "standard oil fluids" at circumstances that do not lead to any categorisation, in enclosed areas does not require any zone classification. Where higher temperatures may be expected (above 50 °C) a Zone 2 classification of the area up to 1 m above any possible spillage pool area is recommended.

Areas above possible spillage pools of category C fluids, i.e. fluids above their T_F, in adequately ventilated areas are recommended to be classified as Zone 2 to a height of 1 m. But in general, localized Zone 1 areas should be classified around open liquid surfaces, surrounded by a larger Zone 2 area.

Note: If hazardous concentrations are created within a building, then there is the potential to produce a hazardous area outside the building, which may lead to a zone classification.

Where an enclosed area is not provided with adequate ventilation, only a more detailed evaluation of the air mixing situation may provide more definite recommendations. In any case, as a starting point the enclosed area should be classified as Zone 1.

8.8 Ambient Temperatures

The IP 15 guidelines are based on a general max ambient temperature of 30 °C unless stated otherwise e.g. regarding zone heights associated with spillage pools.

IP 15 covers oil fluid zone classification in a broader sense, including Class I and Class 0 substances.

The present guidelines focus on what has been defined as "standard oil fluids", which during processing may fall in category C (and no other category). A comparison of the two scopes leads to the conclusion that a general max ambient temperature of 45 °C is appropriate concerning the present guidelines.

Higher general ambient temperatures warrant a more detailed analysis.

The guidelines regarding zone height associated with spillage pools, especially the 50 °C limit, are not affected.

Significant lower ambient temperatures are not foreseen to affect the proposed zone classification.

The steep jump from a 1 to 3 m zone height associated with spillage pools at 50 °C ambient temperature has of course no specific scientific background, and may be regarded as an advice to be implemented in a considered and flexible way.

References

IP 15 "Classification code for installations handling flammable fluids", 3rd edition, July 2005, Publ. by Energy Institute, London

EN 60079-10-1 "Explosive atmospheres—Part 10-1: Classification of areas—Explosive gas atmospheres"

The standard contains an informative Annex D "Flammable mists" with the following two statements:

It is important to point out that not every leak will cause a mist formation, e.g. the leaks through broken flange gaskets or stuffing boxes/packing glands that are the most common secondary grades of release in case of gases or vapours, will usually be negligible in case of viscous liquids and in most cases will cause dripping rather than mist. That means that the likelihood of mists being generated through leaks at pipe joints, valves, etc. should not be overstated.

For liquids released well below their flash point, examples of mist explosions are rare in process industries. This is probably due to difficulty in generating sufficiently small droplets sizes from an accidental release and the associated difficulty of ignition.

EN 1127-1 "Explosion prevention and protection"

Chapter 9
Zone Classification—Natural Gas

9.1 Introduction

Natural gas (methane) is a colourless, odourless, non-toxic (but asphyxiating) highly flammable gas with a composition normally falling in the framework indicated in Table 9.1.

Note: As an average value the methane content may be set to 94 %.

Some physical parameters of natural gas (methane) are presented in Table 9.2.

If natural gas and air is ignited and mixed during the combustion process, the reaction takes the form of a fire.

If natural gas and air is ignited after a sufficient premixing has taken place, the oxidation reaction may proceed much faster, and an explosion may be the outcome.

This chapter sets focus on zone classification around installations handling natural gas. Zoning inside equipment may be relevant for start-up and maintenance. Such periodic zone requirements should take the specific circumstances in consideration, e.g. a possible inerting of the system by using N_2 (Inerting a system is generally equivalent to a one zone step down).

9.2 Grade of Release

In order to designate an appropriate zoning (zone number(s) and extent), the possible sources of release must be identified and assigned a grade of release.

The points, or sources, from which natural gas can be released, have to be identified by careful examination of the plant and its equipment.

The potential for release has to be considered not only in the context of normal operation, but also under conditions of abnormal operation. "Abnormal operation" refers to operation outside the design envelope, and which can be reasonably expected to occur. But **area classification does not cover catastrophic failure**

T. Jespen, *ATEX—Explosive Atmospheres*, Springer Series
in Reliability Engineering, DOI 10.1007/978-3-319-31367-2_9

Table 9.1 General composition of natural gas

Substance	Content (%)
Methane	89–97
Ethane	2–5
Propane	0.1–1.5
Nitrogen	1–5
Carbon dioxide	0.1–1
Hydrogen sulphide (trace amounts)	<5 ppm

Table 9.2 Physical properties of natural gas

Property	Value
Explosive limits	4(5)–15 %
Auto ignition temperature	535 °C
Gas group	II A (T1)
Molecular weight (methane)	16.04
Minimum ignition energy	0.21 mJ
Boiling temperature	−162 °C
Specific gravity (1 atm, 21 °C)	0.55

such as the rupture of vessels, or the complete failure of pipelines, which are beyond the concept of abnormality dealt with in the legally required assessment. However, normal operation includes all conditions within the design parameters of the plant, including start and stop and failure of fragile components.

When a source of release has been identified, it has to be related to one of three grades, continuous, primary, and secondary, as indicated in Table 9.3.

Continuous grades of release will not be dealt with in this chapter, which—with a few exceptions—brings secondary grades of release in focus.

9.2.1 Primary Grade of Release

A primary grade of release is defined as a release which can be expected to occur periodically or occasionally during normal operation.

The following equipment are generally considered as sources of primary grade of release:

- Purge vent pipe terminations, drains and sample points, if operated frequently
- The valve seat of closed purge vent pipes, drains and sample points, if not capped when not in use
- The valve seat of relief valves that are not operating
- Process, machinery and instrument vents, if operating frequently

Table 9.3 Three grades of release

Grade	Definition	Guideline (hours/year)	Associated zone[a]
Continuous	A release which is continuous or is expected to occur frequently or for long periods	t > 1000	0
Primary	A release which can be expected to occur periodically or occasionally during normal operation	10 < t < 1000	1
Secondary	A release which is not expected in normal operation and, if it occurs, is likely to do so only infrequently and for short periods	1 < t < 10	2

[a]Grades are only related to source considerations, while zoning also depends on ventilation circumstances which means that abovementioned general relationship between grade and zone may be deviated

9.2.2 Secondary Grade of Release

A secondary grade of release is defined as a release which is not expected to occur in normal operation and, if it does occur, is likely to do so only infrequently and for short periods.

(a) The following equipment are generally considered as sources of secondary grades of release with a nominal hole size of 0.25 mm^2:

- Pairs of flanges
- Screwed fittings
- Joints
- Distribution regulator diaphragms
- Valve glands

Note: The hole size used for leakage calculations from flange gaskets, screwed fittings and in joints as a metal to metal gap and at valve stem seals is recommended to be taken as 0.25 mm^2 for normal conditions (the fuel is clean, dry natural gas in non-vibratory equipment). For more adverse conditions, a 2.5 mm^2 hole size should be applied.

(b) The following equipment are generally considered as sources of secondary grades of release that are specific to the particular item of equipment and which, in general, are significantly larger than 0.25 mm^2:

- Purge vents, drains and sample points if operating infrequently
- Relief valve vents when operating
- Process, machinery and instrument vents, if operated infrequently
- Shaft seals on compressors and boosters/pumps

9.3 Gas Zones, Definition and Guidelines

Gas zones are defined according to ATEX directive 1999/92 as stated in Table 9.4.

The (universal) guideline presented in Table 9.5 can be used at the zone classification task concerning the duration (t) of the occurrence of an explosive atmosphere.

Note: Zone classification only takes into consideration the type of explosive mixture and the likelihood of its occurrence determined by the frequency and duration of its occurrence. Outside equipment this will be influenced by the degree of ventilation of the area.

Compared with the determination of the **zone extension**, the fixing of the zone type (number) may appear fairly simple. The extension of a natural gas zone will depend on the degree of ventilation of the area and the rate and volume of the release, which is influence by the system pressure and temperature and the release area (e.g. the leakage hole size).

The final determination of zone type and zone extensions are generally based on published examples, experiments or mathematical modelling, which for example underlie recommendations/standards as IGEM/SR/25 (2nd edition).

Table 9.4 Gas zone definitions

Zone	Definition
Zone 0	A place in which an explosive atmosphere consisting of a mixture with air of flammable substances in the form of gas, vapour or mist is present continuously or for long periods or frequently
Zone 1	A place in which an explosive atmosphere consisting of a mixture with air of flammable substances in the form of gas, vapour or mist is likely to occur in normal operation occasionally
Zone 2	A place in which an explosive atmosphere consisting of a mixture with air of flammable substances in the form of gas, vapour or mist is not likely to occur in normal operation but, if it does occur, will persist for a short period only

Table 9.5 Zone classification guided by duration of explosive atmosphere

Zone	Guideline
Zone 0	t > 1000 h/year **or** t > 1 h/shift
Zone 1	10 < t < 1000 h/year
Zone 2	1 < t < 10 h/year **and** t < 1 min/shift

9.4 Effects of Ventilation

Gas or vapour released into the atmosphere can be diluted by dispersion or diffusion into the air until its concentration is below LEL (Lower Explosive Limit). Ventilation, i.e. air movement leading to replacement of the atmosphere around the source of release by fresh air, will promote dispersion, and suitable ventilation rates can also avoid persistence of an explosive gas atmosphere, thus influencing the type of zones.

When considering ventilation, both the degree of ventilation (air movement rate leading to a replacement of the atmosphere by fresh air) and the availability of the ventilation (reliability of it being present) should be taken into account.

EN 60079-10-1 distinguishes three degrees of ventilation, as follows:

- **High ventilation** can reduce the concentration at the source of release virtually instantaneously, resulting in a concentration below LEL. A zone NE (negligible extent) may result.
- **Medium ventilation** can control the concentration, resulting in a stable zone boundary, whilst the release is in progress, and where the explosive gas atmosphere does not persist unduly after the release has stopped.
- **Low ventilation** cannot control the concentration whilst release is in progress and/or cannot prevent undue persistence of a flammable atmosphere after release has stopped.

Three levels of availability of the ventilation are defined in EN 60079-10-1, as follows:

- **Good**: Ventilation is present virtually continuously.
- **Fair**: Ventilation is expected to be present during normal operation. Discontinuities are permitted provided they occur infrequently and for short periods.
- **Poor**: ventilation which does not meet the standard of fair or good, but discontinuities are not expected to occur for long periods.

In order to determine the type of zone, EN 60079-10-1 refers to the following method, which is primarily for use in indoor situations, but is said also to be of use in outdoor situations:

1. Calculate a **hypothetical volume** $\mathbf{V_z}$, see Sect. 9.7
2. Estimate the persistence time of the release, see Annex B, B.5.2.5 of EN 60079-10-1
3. Use Table B.1 of Annex B of EN 60079-10-1 to determine the type of zone

The **hypothetical volume** $\mathbf{V_z}$ represents the volume over which the mean concentration of flammable gas or vapor will typically be either 0.25 or 0.5 times LEL, depending on the choice of safety factor (k) applied to LEL, in order to be on the "safe" side regarding the dilution of the released gas to a concentration below LEL.

Table 9.6 Ventilation rating based on estimated hypothetical volume

Estimated V_z	Ventilation rating
$V_z < 0.1\ m^3$	High ventilation
$V_z > V_0$	Low ventilation
$0.1\ m^3 < V_z < V_0$	Medium ventilation
$0.1\ m^3 < V_z \ll V_0$	Hazardous area as V_z

Table 9.7 Illustration of guidance to zone determination

Grade of release	Ventilation					
	Degree					
	High			Medium		
	Availability					
	Good	Fair	Poor	Good	Fair	Poor
Primary	(Zone 1 NE)	(Zone 1 NE) Zone 2	(Zone 1 NE) Zone 2	Zone 1	Zone 1 + Zone 2	Zone 1 + Zone 2
Secondary	(Zone 2 NE)	(Zone 2 NE)	Zone 2	Zone 2	Zone 2	Zone 2

Notes "Zone 1 NE" (or "Zone 2 NE") stated in brackets, indicates—under the assumption of normal conditions—the existence of a zone 1 (or zone 2) of negligible extent (NE), which means that no action is required to control possible sources of ignition within it. As illustrated above, a "Zone 1 NE" may be surrounded by a "Zone 2"

V_z is said only to be intended to assist in assessing the degree of ventilation, and that V_z is not directly related to the size of the hazardous area.

See Sect. 9.7 for a closer look on the hypothetical volume V_z.

Having estimated V_z, EN 60079-10-1 goes on classifying the degree of ventilation as indicated in Table 9.6 (V_0 is the room volume).

A guide to determine the type of zone is presented in Table B.1 of Annex B to EN 60079-10-1. Table 9.7 illustrates a part of this Table B.1.

A **zone of negligible extent** can be seen as the influence of explosion risk assessment on zone classification (see also Sect. 9.6) in areas where the ventilation can be regarded as "high" relative to the leak size.

In some (normal) situations a released cloud can be so small and contains so little flammable material that its ignition is an insignificant event and will cause no injury to persons in the vicinity or damage to equipment.

As an example, small enclosures, say below 10 m^3, are outside the range of normal conditions as the possible explosion pressure under such circumstances cannot be regarded as insignificant.

"+" signifies "surrounded by".

The Zone 2 area created by secondary grade of release may exceed that attributable to a primary grade of release; in this case, the greater distance should be taken.

More about the use of V_z in zone type estimation, including criticism and newer developments, to be found in Sect. 9.7.

9.5 Natural Gas Zone Classification Around Installations

The IGEM/SR/25 standard applies to liquid-free natural gas with a combined proportion of methane-plus-inerts of greater than 89 % by volume; a gross calorific value not exceeding 45 MJ/m^3; molecular weight not exceeding 20 kg/kmol; LEL of not less than 4.4 %; a temperature range of −20 to 50 °C and pressure not exceeding 100 barg. Ambient temperatures are assumed to be in the range −20 to 35 °C.

NE (negligible extent) zone classification is restricted to operating pressures below 10 barg, and a max release of 1 g/s in a freely ventilated location.

9.5.1 Leakage

Zoning due to possible leakage from "Screwed fittings and joints up to 50 mm nominal size", "Flanges", "Valve connections", "Valve glands", "Regulator components including breather" and "Regulator components including breather vent pipes" is illustrated in Fig. 1 a–f in IGEM/SR/25, with the zoning distance (X) to be found in Table 1 and 2 of the publication.

Note: Leakage zoning distances are allowed to overlap without interference considerations (opposite what may have to be considered regarding vent pipe terminations that may relieve simultaneously, or near simultaneously).

9.5.2 Non-ideal Vent Pipe Terminations Associated with Relief Valves

Vent pipe terminations associated with relief valves tend to give rise to zone 2, as relief valves usually are operated infrequently, but may be subject to small valve seat leakage.

"Non-ideal" signifies that this design in general not give rise to the shortest zone distance (when compared to impeded or upward pointing and angled designs), but may be the best and preferred design for vent terminations in some situations.

As an example, zoning related to **non ideal vents downwards pointing** is illustrated in Fig. 16 of IGEM/SR/25.

Zoning regarding vents associated with process machinery, continuous flowing instrumentation, other vent designs and other flow rates, vent pipe diameters and heights can also be found in IGEM/SR/25.

The vent pipe **flow rate** is an important parameter for the zoning task and should be obtained from the manufacturer or the design data for the equipment connected to the pipe.

Where such information is not available, it may be possible to calculate the flow rate more or less sophisticated, but the equations for flow rate calculations stated in Sect. 9.7 are considered to be of sufficient accuracy.

Section 9.7 contains information on the relationship between operating pressure and estimated mass flow rates (G) for hole sizes of 0.25 and 2.5 mm^2, and on the relationship between mass flow rates (G_m) and volume flow rates (G_v).

Note: If two or more vents release flammable material simultaneously, there is a possibility of released gases (rather than dilution air) being induced from one vent into another. If this happens, the subsequent dispersion to a safe distance will be greater that would have occurred had the vents been considered individually.

Appendix 8 of IGEM/SR/25 deals with the determination of the composite hazardous area for multiple vent stacks.

9.6 Zone Classification and Risk Assessment

Zone classification is about assessing/estimating the likelihood that explosive atmospheres will occur, their type and persistence and the extension of the hazardous area.

Zone classification is a key element of the explosion risk assessment obligation laid down on the plant owner (the employer) forming part of his "Explosion protection document", which must be drawn up prior to the commencement of work. The risk assessment obligation also includes an estimate of the likelihood that ignition sources will be present possessing sufficient energy to ignite the explosive atmosphere, and the scale of the anticipated associated consequences,

As mentioned in EN 60079-10-1 (clause 4.2), a specific zone classification is to be supplemented by a risk assessment in order to decide about the basic equipment selection rules (combination of zone type and equipment category) ought to be deviated.

Zone classification is not an exact science, and in some situations it will be appropriate to operate on a higher "safety" level (choosing a stricter zone classification and/or the use of more safe equipment) or to use less safe equipment than specified by the legislation—based on sound risk assessment considerations.

9.7 Adequacy of Ventilation

9.7.1 Assessing the Adequacy of Ventilation

EN 60079-10-1 provides a methodology for assessing the degree of ventilation for a given release rate G (kg/s) and air change rate C (s^{-1}), based on the calculation/estimation of a key parameter V_z (the hypothetical volume). The parameter represents (is defined as) the volume over which the mean concentration of the released flammable gas will be "k" times the LEL_m:

$$V_z = (f/C) \times G/(k \times LEL_m) \times T/293$$

where:

Factor/symbol	Explanation
f	Correction factor to account for non-homogeneous and non-instantaneous mixing at the source, ranging from 1 (ideal situation) to 5 (impeded airflow)
C	Fresh air changes per unit time (s^{-1}) As a very conservatives approximation, EN 60079-10-1 uses a C = 0.03 s^{-1} (108 h^{-1}) for an open air situation (The focus of the equation is the in-door situation)
k	Safety factor (applied to LEL), typically 0.25 for continuous and primary grades of release, and 0.5 for secondary grades of release
LEL_m	LEL (Lower Explosive Limit) in mass per volume (kg/m^3) $LEL_m = 4.16 \times 10^{-4} \times M \times LEL_v \times 293/T$
LEL_v	LEL in volume per volume (vol.%)
M	Molecular mass (kg/kmol)
T	Temperature (K)
G	Max release rate in mass per time (kg/s)

EN 60079-10-1 provides (in its Annex A) rather complicated equations for estimating the release rate, but the following more simple equations put forward in IGEM/SR/25, give a sufficient accuracy (performed test calculations differ only insignificantly):

1. Release of gas with **choked velocity** (sonic release), i.e. a velocity equal to the speed of sound for the gas, which for natural gas (methane) applies at operating pressures (P) above 0.85 barg:

$$G = 675 \times C_d \times A \times M^{0.5} \times T^{-0.5} \times (P + 1.013)^{1.05}$$

2. Release of gas with **non-choked velocity** (subsonic release), i.e. a velocity below the speed of sound for the gas, which for natural gas (methane) applies at operating pressures (P) below 0.85 barg:

$$G = 1500 \times C_d \times A \times M^{0.5} \times T^{-0.5} \times P^{0.5}$$

where:

G Release mass flow rate (kg/s)
P Operating pressure (barg)
M Molecular weight (kg/kmol)

T Temperature (K)
A Cross-sectional area of the opening (m^2)
C_d Discharge coefficient (For orifices in relief valves: Manufacturer's value or 1, and for other orifices: 0.8 or lower in case of specific knowledge)

9.7.2 Degree of Ventilation

EN 60079-10-1 uses the following system to evaluate the degree of ventilation of an area/volume (V_0) served by the actual ventilation in the vicinity of the release being considered (For in-door situations, V_0 will generally be the volume of the room or building being considered unless there is ventilation specific and local to the release being considered):

High ventilation (HV), when $V_z \leq 0.1\ m^3$ or less than 1 % of V_o—whichever is the smaller.

Fulfilling the HV criteria is a basic requirement for the hazardous area to be classified as a Zone 2 NE (Negligible Extent), which means a Zone 2 area having so small a volume that no action is required to control sources of ignition within it (see also subsequent text).

Low ventilation (LV), when $V_z > V_o$ (volume of the enclosure—see above).

Medium ventilation (MV), when the ventilation is neither HV or LV.

9.7.3 Setting Focus on Natural Gas (Methane)

In the literature you meet slightly different basic data for methane and natural gas because natural gas also contains minor amounts of other gasses:

Methane	M: 16 kg/kmol	LEL_v = 5 %	LEL_m = 0.033 kg/m^3 (20 °C)
Natural gas	M: 18.5 kg/kmol	LEL_v = 4.4 %	LEL_m = 0.034 kg/m^3 (20 °C)

The equation for calculating (estimating) V_z can at a temperature of 20 °C be simplified to (using a value of C_d = 1 and k = 0.5):

$$V_z = f/C \times G/(0.5 \times 0.03) = f/C \times G/0.015 (m^3)$$

Choked release then to be estimated by:

$$\begin{aligned} G &= 675 \times 1 \times A \times (16/293)^{0.5} \times (P + 1.013)^{1.05} \\ &= 158 \times A \times (P + 1.013)^{1.05} (kg/s) \end{aligned}$$

Subsonic release then to be estimated by:

$$G = 1500 \times 1 \times A \times (16 \times P/293)^{0.5}$$
$$= 351 \times A \times P^{0.5} (kg/s)$$

An example:

At an operating pressure P = 10 barg and an estimated hole size of 0.25 mm^2, the release rate to be estimated by:

$$G = 158 \times 0.25 \times 10^{-6} \times (10 + 1.013)^{1.05} = 0.49 \times 10^{-3}\,kg/s = 0.49\,g/s$$

That leads to an estimated V_z volume in an ideal open air situation:

$$V_z = 1/0.03 \times 0.49 \times 10^{-3}/0.015 = 1.1\,m^3$$

using a C = 0.03 s^{-1} (=108 h^{-1}) value for open air as put forward by EN 60079-10-1.

In an enclosure (room) having, what is generally defined as "adequate ventilation", a C = 12 h^{-1} (=0.003 s^{-1}), the V_z volume will be estimated to:

V_z = 11 m^3 (10 times the value for open air) or

V_z = 22 m^3, if a little more conservative f = 2 were applied.

Note: It is worth noting that the abovementioned equation for estimating V_z—together with some other items—are not really a part of the standard EN 60079-10-1, but put forward in informative parts (annexes) of the standard.

As illustrated by the previous calculations, a low V_z value requires very low release rates or very high degrees of ventilation (and the equation also appears somewhat "hypothetical" for open air considerations having a parameter as the number of air changes included).

Studies using Computational Fluid Dynamics (CFD) as RR630, have revealed a very significant conservative aspect of the V_z calculation/estimation according to EN 60079-10-1.

As an example, a 5.2 g/s methane release leads to a V_z between 12.3 and 61.5 m^3, depending on appropriate ventilation quality factor f (1 to 5), while a CFD calculation gives a V_z = 0.09 m^3. The EN 60079-10-1 equation may therefore overestimate V_z by nearly three orders of magnitude.

Based on the negligence of precautionary measures for $V_z \leq 0.1$ m^3 (and the original definition of V_z), RR630 recommends a max leak rate (G) of:

- 2 g/s for unobstructed open air situations
- 1 g/s for obstructed open air situations
- 1 g/s (absolute max) for indoor (enclosures) situations

in order for the classification of an area as Zone 2 NE.

IGEM/SR/25 recognizes the concept of V_z ($\leq 0.1\ m^3$) as the basis for a Zone 2 NE classification, but requires the max operating pressure not to exceed 10 barg and the calculated max release rate (G) not to exceed 1 g/s, and the release to be located in a freely ventilated location.

In drafts aiming at updating EN 60079-10-1, one finds new (and complicated) equations (now separate for outdoor and indoor releases) for estimating V_z.

As an example: For an indoor release of natural gas at 10 barg through a 0.25 mm^2 hole into a 45 m^3 well ventilated enclosure, the calculations (according to updating drafts of EN 60079-10-1) become:

$$G = 0.47\,g/s \quad \text{and} \quad V_z = 0.0128\,m^3$$

A CFD model predicts V_z = 0.0039 m^3 which underlines the still conservative approach put down by the EN 60079-10-1 calculations.

9.7.4 Supplementary Natural Gas Release Calculations

In order to supplement the information in Sect. 9.5 concerning the zoning distance tables to be found in the literature, Table 9.8 illustrates the relationship, that exists between operation pressure and release at hole sizes of 0.25 and 2.5 mm^2 for **choked releases**, i.e. for operating pressures above 0.85 barg.

At 20 °C (293 K) the release rate G_m in g/s can be stated in m^3/h as indicated in Table 9.9.

Table 9.8 Relationship between pressure and two hole sizes

Pressure (barg)	A = 0.25 mm^2	A = 2.5 mm^2
	G (g/s)	G (g/s)
2	0.126	1.26
5	0.260	2.60
7	0.351	3.51
10	0.490	4.90
20	0.967	9.67

Table 9.9 Release rate relationship

Release rate G_m (g/s)	Release rate G_v (m^3/h)
0.05	0.25
0.2	1
2	10
5	25
10	50
20	100

Values at other temperatures can be calculated by using the ideal gas law:

$$\begin{aligned}\text{Density: } \rho(\text{kg/m}^3) &= n \times M/V = P \times M/(R \times T)\\ &= M/(0.0821 \times T) \text{ at 1 atm}\\ &= 211/T\end{aligned}$$

References

IGEM/SR/25, Edition 2, Communication 1748 "Hazardous area classification of natural gas installations", 2010

EN 60079-10-1 "Explosive atmospheres - Part 10-1: Classification of areas - Explosive gas atmospheres"

RR630 Research Report "Area classification for secondary releases from low pressure natural gas systems", HSE, 2008

Chapter 10
Zone Classification—Dusts

Abbreviations

D	Dust
G	Gas
IP code	Ingress protection code
K_{st}	Rate of pressure increase in a 1 m^3 volume (bar × m × s^{-1})
LEL	Lower explosive limit equivalent to lower flammable limit and minimum explosive concentration
LIT	Layer ignition temperature of dust layer
MIE	Minimum ignition energy
MIT	Minimum ignition temperature of dust cloud
P_{max}	Max explosion pressure
T_{ig}	See MIT
T_l	See LIT

10.1 Introduction

Most solid organic materials, as well as many metals and some non-metallic inorganic materials can burn or explode if finely divided and dispersed in air in sufficient concentrations.

Two major differences exist between zone classification for gas/vapour and dust, which are concerned with the spread of flammable atmospheres and the ignition characteristics.

When a gas/vapour is released, the resultant flammable cloud will gradually disperse into the atmosphere. When the gas/vapour concentration has fallen below the lower explosive limit (LEL) there is then no ignition risk. This helps to define zone boundaries with some precision.

In contrast, the release of dust will form a cloud initially, which will settle on to the floor, ledges, etc., from where it could be stirred up again to form a flammable cloud. This means that a potential risk remains, and this has to be taken into account

T. Jespen, *ATEX—Explosive Atmospheres*, Springer Series
in Reliability Engineering, DOI 10.1007/978-3-319-31367-2_10

in area classification decisions. 1 mm layer of dust having a density of 500 kg/m^3 may create a dust cloud of 100 g/m^3 to a height of 5 m.

It is not really possible to calculate the extents of hazardous zones outside equipment for dust as for gases and vapours.

Flammable gas/vapour clouds are sensitive to ignition by low energy sources, but their auto-ignition temperatures are generally in excess of 250 °C, with the majority being above 35 °C. It is very unlikely, therefore, that the outer surface of electrical apparatus will get sufficiently hot to cause direct ignition of a flammable gas/vapour/mist cloud. Surface temperatures must, of course, be taken into account when selecting apparatus.

Many powders have dust cloud ignition temperatures that are significantly lower than the auto-ignition temperatures of gases/vapours. In addition to this, layers of powder on hot surfaces will often decompose when exposed to temperatures as low as 150–200 °C. This decomposition can lead to smouldering or combustion with flame, producing a potential source of ignition for any flammable dust cloud in the vicinity. This is another aspect that has to be considered in the selection of apparatus.

A lot of information may be required to carry out a **zone classification** for a potential explosive dust atmosphere (i.e. the estimation of the likelihood of its occurrence) supplemented by an **explosion risk assessment**, which also includes an estimation of the likelihood of effective ignition sources being present and of the anticipated consequences of an explosion. The risk assessment covering the specific circumstances may indicate an appropriate derogation from the basic relationship between zone type and equipment selection, as stated in ATEX Workplace directive 1999/92.

10.2 Explosion Risk Parameters

Besides a detailed knowledge of the operating procedures, the following properties of the material should be available, as appropriate:

- Fuel classification as **Group A** (explosive) or **Group B** (not flammable at ambient temperatures—i.e. up to 110 °C) and/or flammable at elevated temperatures, is the first (basic) step in dust flammability considerations.
- **Physical properties**, such as particle size and size distribution, particle shape, particle surface condition and moisture content: Where the particles exceed diameters of about 0.5 mm, flammable dust clouds will not be formed. Maximum ignitability occurs with particle sizes of 0.075 mm or smaller and material of this fineness should always be used for tests. Reducing the particle size will reduce (often very significantly) the necessary minimum ignition energy (MIE). Lowering the moisture content will also reduce the MIE.
- **Minimum explosive concentration** (LEL). Similar to gas/vapour, dust in the form of a cloud is only ignitable over a range of concentrations, but only the

lower limit (LEL) is of practical use, because of the inherent difficulties in achieving homogeneous dust clouds at high concentrations. The LEL for most dusts is in the range 20–60 g/m^3, but some materials have values much higher than this, reaching 250–300 g/m^3. Temperature and pressure will influence the limits of flammability for dusts in a similar way to those for gas/vapour, but this is of little practical significance as the Upper Explosive Limit (UEL) is not used, and the changes in the small concentrations at the LEL will usually have little effects on the hazards.

- **Minimum Ignition Energy** (MIE), is an important parameter in explosibility considerations—not directly for zone classification but for the supplementary explosion risk assessment—see further comments in Sect. 10.5.1.
- **Minimum Ignition Temperature** (MIT of a dust cloud), is an important parameter for the explosion risk assessment and equipment selection. MIT does not enter zone classification deliberations.
- **Layer Ignition Temperature** (LIT of a dust layer), is an important parameter for explosion risk assessment and equipment selection. LIT does not enter zone classification deliberations. LIT varies with the material being tested, and generally falls as the thickness of the layer increases. Tests are usually carried out on 5 mm thick layers, as layers in hazardous areas should not reach this depth. The ignition temperatures obtained from the test can be influenced by the thermal prehistory of the material, so that it is important to ensure that the sample submitted for testing is representative (A circumstance of general importance when determining explosion parameters for dust, which means that it may be a better idea to use "standard" values instead).
- **Bulk powder volume resistivity**, which is not a flammability property, but may be of importance for equipment selection as a conducting layer of dust could cause short-circuits within electrical equipment—see further comments in Sect. 10.5.2.

The concept of **Flash point** that is used to indicate flammability in vapours is not appropriate to dusts.

Note: The possible release of flammable gases contained in the solid material may be an important aspect to include in the explosion risk considerations, cf. Sect. 10.8.4 on methane in coal.

10.3 Grade of Release

Areas or points in the system where dust clouds can occur, or layers may accumulate are identified as **sources of release**. The sources are graded as "continuous", "primary" or "secondary", in accordance with the frequency with which dust clouds may be formed—in line with the similar assignment regarding gases and vapours.

A "continuous" source will almost always be inside plant equipment, and a "secondary" source will apply primarily to locations where dust clouds may occur from the disturbance of layers only (e.g. at reloading of dry materials from belt conveyors).

A continuous source of release which creates dust clouds outside equipment is not acceptable, and steps must be taken to remove it.

More about similarities and differences between "Grade of release" and "zone classification" can be found in Chap. 2.

Table 10.1 relates the type of source to the appropriate zone, and indicates time bands that could help in identifying the grade of release.

10.4 Zone and Zone Extent Assignment

In common with the procedures used with gases and vapors, the zone definitions used for combustible dusts shall reflect the likelihood with which flammable dust atmospheres will be formed in the plant. But they must also take account of the influence of the combustion characteristics of dusts, and the presence of layers (Dust layers outside buildings are very unlikely to form flammable dust clouds).

The dust zone types and definitions are similar to those for gases and vapours, as indicated in Table 10.2.

Table 10.3 can be used as an interpretation of the time related statements of Table 10.2.

The **extents of zones** are affected by operating procedures, ventilation (unless local exhaust ventilation is near the source of release, ventilation will not affect zone extent significantly), housekeeping standards, etc. Good housekeeping, appropriately monitored, can modify a possible zone 22 area into being a non-hazardous area.

Guidance on housekeeping can be found in EN 60079-10-2, Annex C.

Concerning zones and their extents, the following guidelines may be used.

Zone 20
Zone 20 locations will be inside equipment, and will be defined by the walls of the various items.

Table 10.1 Specification on different Grades of release

Grade	Definition	Guideline (hours/year)	Associated zone
Continuous	Release which is continuous or is expected to occur frequently or for long periods	$t > 1000$	20
Primary	Release which can be expected to occur periodically or occasionally during normal operation	$10 < t < 1000$	21
Secondary	Release which is not expected to occur in normal operation and, if it does occur, is likely to do so only infrequently and for short periods	$1 < t < 10$	22

Table 10.2 Dust zone classifications/definitions

Zone	Definition
Zone 20	A place in which an explosive atmosphere in the form of a cloud of combustible dust in air is present continuously, or for long periods or frequently
Zone 21	A place in which an explosive atmosphere in the form of a cloud of combustible dust in air is likely to occur in normal operation occasionally
Zone 22	A place in which an explosive atmosphere in the form of a cloud of combustible dust in air is not likely to occur in normal operation but, if it does occur, will persist for a short period only

Table 10.3 Guidance on time in relation to zone classification

Zone	Guideline
Zone 20	t > 1000 h/year **or** t > 1 h/shift
Zone 21	10 < t < 1000 h/year
Zone 22	1 < t < 10 h/year **and** t < 1 min/shift

Zone 21

Zone 21 areas can be expected to occur around covers removed routinely for operating purposes, areas near to bag or drum filling or emptying points, around weighing out stations and sampling points, etc.

Zone 21 will normally be of only a small extent, and will rarely exceed 1 m from the nearest point of a source of release; extending vertically downwards to the ground or a solid floor, and to 1 m above the point of release.

Where the spread of dust is limited by mechanical structures (walls, etc.), their surfaces can be taken as the boundary of the zone.

A non-confined zone 21 (not limited by mechanical structures, e.g. a vessel with an open man-hole) located inside, will usually be surrounded by a zone 22.

Zone 21 can also be present inside some kind of equipment, such as the coal mill, separator, ducts, silos and feeders in a coal firing system.

Zone 22

Typically, the extension of a zone 22 will be no more than 1 m from a possible release source.

Zone 22 is generally associated with outlets from cyclones or bag filters where the clean side of the filter may contain a flammable dust cloud if the filter element fails. This applies to the interior of the filter, downstream of the filter element, extract ducting and around the discharge of the extract duct. A zone 22 is also of relevance regarding pneumatic systems where malfunction can lead to dust release, equipment that is opened on rare occasions for cleaning, etc., as well as areas where deposited dust can be present.

The extent of zone 22, which is essentially where layers could be formed, should be as small as practicable, e.g. 3 m beyond the zone 21.

In practical terms it is best to estimate the extent of zones 22 by comparison with powder spread in units employing the same equipment with similar standards of housekeeping, ventilation, etc.

Alternatively, useful guidance can often be gained from experience of other people with the equipment, or different equipment in similar situations. Where there is no precedent it is necessary to define the level of housekeeping, and ventilation, and make reasonable estimates of dust spread, taking account of working methods, and the fact that in some cases a flammable dust cloud could be formed from a layer only 1 mm thick.

Working methods can have a major effect on the size of zone 22. For example, a larger zone could be needed where powder is emptied from bags or flexible intermediate bulk containers, than where it is emptied from drums or rigid intermediate bulk containers because of the greater risk of spillage, and the need for flattening and disposal of the bags.

Physical barriers such as walls, partitions, etc., can be used to define zone boundaries with great advantage.

10.5 Explosion Risk Assessment Aspects

The starting of an explosion requires:

1. A flammable substance(s) and
2. Oxygen (air) which are to be
3. Premixed in the right proportions (concentrations within explosive limits), and
4. A reasonable high degree of confinement, and
5. An ignition source of sufficient energy.

Note: The scope of the ATEX directives is potential explosive atmospheres, which are defined as mixtures with air under atmospheric conditions of flammable substances, in which, after ignition has occurred, combustion spreads to the entire unburned mixture. Atmosphere conditions are regarded as pressures between 0.8 and 1.1 bar, and temperatures between −20 and 60 °C.

Where relevant, risk assessments and coping with conditions outside the area of "atmospheric conditions" must of course not be ignored, but such situations are not directly regulated under the ATEX directives (which however may provide inspiration for tackling these risks).

Particle size has a profound effect on the reaction (oxidation) rate due to surface area available for reaction: Reduction to 1/8th of the original particle size results in an 800 % increase in surface area.

Reducing the particle size has also a significant influence on the required MIE.

Moisture in dust reduces both ignition sensitivity and explosion violence; firstly by minimising dispersibility of the dust to form a cloud and secondly by absorbing heat during combustion. As a general rule, explosibility requires moisture content below 11 %.

LEL, Lower Explosive (or flammability) Limit—or minimum concentration for explosibility—will vary with the type of dust, but a value of about 40 g/m^3 is generally appropriate for the risk assessments. The Upper Explosive Limit (UEL) is not of any practicable value for dust fuels.

Oxygen concentrations above 21 % (v/v) tend to enhance combustion, and lower oxygen concentrations in the range 9–15 % (v/v) will create non-flammable conditions.

Explosive oxidation reactions create rapid high temperature increase which causes pressure increases. Both max explosion pressure (P_{max}) and the rate (K_{st}) with which the pressure increases, is of importance in explosion risk assessments. Because the rate of pressure rise is volume dependent, the volume-independent constant K_{st} (K_g for gases) is used: K_{st} = dP/dt in a 1 m^3 volume (unit: bar × m × s^{-1}).

The rate of pressure rise is often stated as a class as stated in Table 10.4 (Table 10.5).

Explosion relevant parameters are very much dependent on especially dust size and distribution, and naturally the moisture content, so the literature-data presented in Tables 10.6 and 10.7 should only be regarded as illustrative examples.

T_{ig} and T_l are used in **equipment selection** to specify the maximum surface temperature (T_{max}) of the equipment to be placed in the zone in the following way in order to have an appropriate margin of safety.

T_{max} must comply with **both** of the following criteria:

1. $T_{max} \leq 2/3 \times T_{ig}$ (i.e. 2/3 × 430 = 286 °C for lignite coal dust), and
2. $T_{max} \leq T_l - 75$ (i.e. 230 − 75 = 155 °C for lignite coal dust)

If a layer of dust more than 5 mm thick is expected, the maximum permissible equipment surface temperature (T_{max}) should be reduced.

T_{max} should be stated in the ATEX marking of equipment for use in ATEX dust zones (Regarding equipment for use in ATEX gas zones the equipment max temperature is indicated in the ATEX marking by the use of temperature classes: T1 to T6).

Table 10.4 St classes

St class	K_{st} (bar m s^{-1})	Specification
St 1	<200	Weak explosion
St 2	201–300	Strong explosion
St 3	>300	Very strong explosion

Table 10.5 K_{st} for some dust types (examples from the literature)

Dust	K_{st} (bar m s^{-1})
Flour	87
Starch	150
Aspirin	217
Pigment	286
Aluminium	555

Table 10.6 Explosibility parameters—non-coal dust

Dust type	LEL (g/m³)	P_{max} (bar)	K_{st} (bar m s^{-1})	T_{ig} (°C)	T_l (°C)
Wood dust	30–60	8.9–10.5	130–190	420–520	310
Rice	60	6.7	91	510	450
Wheat flour	60–125	8.0–9.3	139	440	440
Sugar	30	9.0	123	370	400
Cellulose	125	8.9	56	480	270

Table 10.7 Explosibility parameters—coal dust

Coal dust	LEL (g/m³)	P_{max} (bar)	K_{st} (bar m s^{-1})	T_{ig} (°C)	T_l (°C)
Lignite (brown coal) (VC > 36 %)	60	9.3	110–175	430	230
Bituminous—high volatile (26 < VC ≤ 36)	30–60	9.1	70–120	550–580	240
Bituminous—low volatile (11 < VC ≤ 26)	30–60	8.3	80–90	600–690	260–285
Pet coke (8 < VC ≤ 11)	30–125	6.9–7.2	45	690	280–320
Anthracite (VC ≤ 8)	100	0.6	1–2	700	340

LEL Lower explosive limit
P_{max} Max explosion pressure
K_{st} "Explosion constant"—max rate of pressure rise (see above)
T_{ig} Minimum dust cloud ignition temperature
T_l Minimum dust layer (5 mm) ignition temperature (also indicated by T_s)
VC Volatile content (%)

10.5.1 Minimum Ignition Energy (MIE)

Minimum Ignition Energy (MIE) aspects concerning dust are dealt with in more detail in Chap. 3, Sect. 3.3.3.

The Minimum Ignition Energy (MIE) is a measure of spark sensitivity—used for assessing static and mechanical ignition sources. MIE is affected by many factors including: Plant conditions as temperature, pressure, turbulence and oxygen concentration, powder properties as particle size, exact chemical composition and moisture content, and test conditions as electrode details and discharge circuit.

MIE of dust clouds varies over at least eight orders of magnitude, from about 0.01 mJ for very fine sulphur powder to beyond 1000 J at the upper end.

MIE for dust is often reported to fall between 10 and 100 mJ (or above), but values down around 1 mJ and significantly higher values are also seen in the literature. MIEs for dust are reduced significantly at elevated temperatures and by the presence of flammable gas/vapour, even below the LEL.

MIEs for dusts are not used in area classification, neither the specification of apparatus. Their primary use is in the control of electrostatic hazards (With MIE above 500 mJ, the dust cloud is normally considered as being non-sensitive to ignition by electrostatic discharge).

Where an exact MIE value is of importance for explosion risk assessments, a representative sample of the relevant dust has to be investigated (tested).

Examples of MIE values taken from the **literature** are listed in Chap. 3, Sect. 3.3.3.

Figure 10.1 (reproduced from Chap. 3, Sect. 3.3.3) conveys some aspects between MIE energies for gases and dusts and possible ignition sources.

10.5.2 Powder Volume Resistivity

Powder volume resistivity aspects are dealt with in more detail in Chap. 3, Sect. 3.3.8.

The Powder Volume Resistivity (Ω × m) is a measure of the resistivity through the bulk of the powder and indicates the ability of the powder to retain charge. It is

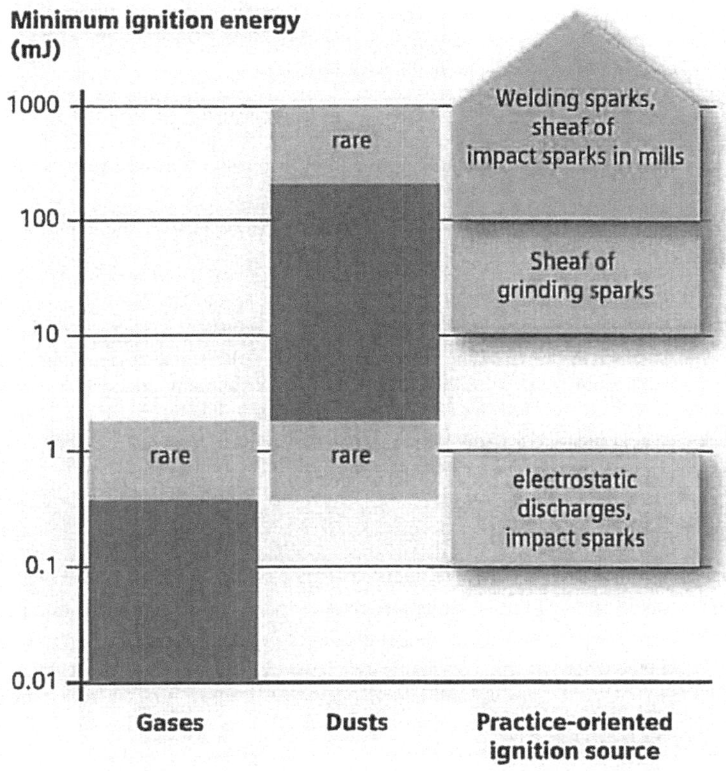

Fig. 10.1 MIE energies for gases and dusts

used to assess the likelihood of accumulating electrostatic charge on the bulk powder, especially when the MIE is low, say below 25 mJ.

The volume resistivity (ρ) in unit "Ω m" of a specimen having an area A (m^2) and a length l (m) is determined by the equation:

$$\mathrm{R(Resistance)} = \rho \times (l/A)\,\Omega\ (\mathrm{Ohm})$$

Volume resistivity covers a very broad scale from about 10^{-8} Ω m for metals to about 10^{16} Ω m for super insulators.

Resistivity values taken from the **literature** are listed in Chap. 3, Sect. 3.3.8.

EN 60079-20-2 (If/when published) defines, in continuation of the specifications in EN 60079-0, three dust groups (capable of forming explosive atmospheres) with the resistivity as one of the parameters, as illustrated in Table 10.8.

In North America explosive dust is subdivided into the three material groups as presented in Table 10.9.

If the volume resistivity is $>10^{12}$ Ω m, then the powder is regarded as insulating. Powders with a volume resistivity $<10^{9}$ Ω m are considered to have low to medium resistivity. It is rare for powders to have resistivity much lower than this. Even

Table 10.8 Possible future specifications of three dust groups

Dust group	Type	Definition	Examples
IIIA	Combustible flyings	Solid particles, including fibres, <500 µm Can burn or glow in air	Rayon, cotton, sisal, jute, hemp, oakum, waste kapok
IIIB	Non-conductive combustible dust	Fine solid particles, ≤500 µm Resistivity $>10^3$ Ω m	Corn, custard powder, cellulose acetate, nylon, sugar, polythene, epoxy
IIIC	Conductive combustible dust	Fine solid particles, ≤500 µm Resistivity $\leq 10^3$ Ω m	Aluminium, magnesium, tin, cadmium, coal dust[a]

[a]Not all types of coal dust are to be regarded as conductive dust

Table 10.9 Explosive dust groups specified in North America

Dust group	Specification	Examples
E	Metal dust (explosive and conductive)	Aluminium, magnesium
F[a]	Carbonaceous dust (explosive and some conductive)	Carbon black, coal dust, coke dust
G	Grain dust (explosive and non-conductive)	Flour, sugar, spices, certain polymers

[a]Group F is often specified with the requirement of >8 % entrapped volatiles

metal powders can have significant resistivity as an insulating oxide film builds up on the surface of them.

Bulk powders can be split into three volume resistivity categories:

- Low resistivity: $<10^6\ \Omega$ m
- Medium resistivity: 10^6 to $10^{10}\ \Omega$ m
- High resistivity: $>10^{10}\ \Omega$ m

In practice, low resistivity powders are rare because metal powders will eventually oxidize.

When Powder Volume Resistivity is $>10^9\ \Omega$ m, there is no electrostatic risk in earthed metal plants provided that Minimum Ignition Energy is >10 mJ.

10.6 Controlling Dust Explosion Risks

Note: More information can be found in Chaps. 3, 4, 5 and 6.

The ATEX directives presuppose compliance with integrated explosion safety, i.e. in order of priority:

- Preventing the formation of explosive dust atmosphere
- Preventing sources of ignition
- Mitigating the effects of an explosion where the possible occurrence must be accepted

Preventing the formation of dust clouds corresponds to avoiding that areas have to be zone classified, e.g. by avoiding dust accumulations, which subsequently can become airborne. This may be achieved by evading areas where dust may settle (promoted by effective housekeeping) or by maintaining a low concentration of oxygen (O_2) through the partial or complete substitution of the air by an inert gas, as nitrogen (N_2) or carbon dioxide (CO_2).

Explosive dust can become non-explosive by diluting them with inert dust, as calcium sulphate, limestone, sodium bicarbonate, common salt, volcanic ash (trass), various silicates or stone dust. These materials may act as a heat sink or otherwise interfere with flame propagation. In general, at least 60 % diluents dust is required, and it is necessary for the diluents to be intimately mixed with the explosive dust.

Dust accumulations can be minimized by maintaining transport speed above about 20 m/s in pipe lines and ducts, avoiding horizontal surfaces and aiming at using walls with smooth surfaces

Outside equipment, the most important anti-explosion measure—apart from placing the equipment in open air—is maintaining the process buildings in a clean condition. Accumulated dust deposits can provide fuel for secondary explosions. Indoor, the preferred method is a vacuum system rather than brushes and shovels which create dust clouds.

The first step at preventing dust deposits within a building is of course to maintain a plant as leak-tight as possible, or alternatively to operate at a slightly lower pressure than atmospheric pressure to reduce the escape of dust.

Avoiding effective ignition sources stemming from equipment, is primarily a question about using, installing and maintaining appropriate (categorized and Ex-marked according to zone requirements) equipment in zone classified areas.

The possibility of ignition due to hot surfaces, hot work and the most frequently experienced dust explosion sources: mechanical friction/heating and smouldering must always be in focus at risk assessment scenarios and elaboration of instruction materials. The most frequently registered general mechanical ignition sources (registered by equipment type) are: Polishing, mill/size reduction, conveying, mixer, dust collector and silo/bunker.

Impact sparks are likely to arise where tramp metal or stones enter the process plant. A magnetic separator to catch ferrous tramp metal is a very widely used precaution that helps minimise this problem.

In some cases the possibility of an explosion has to be realised and accepted (e.g. because of the material or impurities of the material being processed), and you have to **implement mitigating measures** to limit the effects of an explosion to an acceptable risk level.

Mitigation measures fall into the following main categories:

- Explosion relief venting (most widely used technique for protecting the process plant)
- Explosion suppression
- Prevention of flame and explosion propagation
- Explosion containment (when the equipment is not too big)

Explosion relief venting is brought about by providing the equipment with one or more deliberate points of weakness (named "explosion relief vents") with the intention to prevent injuries to persons nearby by avoiding uncontrolled failure of the equipment.

When an explosion vent opens as a result of a dust explosion, a fireball or jet of flame must be expected. Therefore, explosion vents inside buildings are often fitted with a duct (or a deflector) to lead the explosion products to a safe place in the open air.

Explosion vents must open reliably at a pressure well below that which the plant equipment, it is protecting, can withstand.

Explosion suppression is often used where it is not possible to protect by using containment or explosion relief venting, and in particular where the pressure and flame of the explosion cannot be vented to a safe location (e.g. caused by its content of toxic dust or other harmful substances).

Explosion suppression requires the incipient explosion to be detected very soon after ignition, and that sufficient suppressant can be discharged into the growing fireball in the enclosure at a fast enough rate to extinguish all flame before a destructive overpressure develops.

The separate items of a dust-handling plant are usually connected together, e.g. by ductwork, chutes, pipes and conveyors, and an explosion initiated in one item can therefore propagate along these lines and start a subsequent explosion in other plant item.

Various approaches and techniques can be used to **counteract explosion propagations**. As an example, explosion diverters in ducts and screw conveyors can provide "a choke" of material to prevent the propagation of an explosion, but the appropriate choice of method/technique must always rest on a systematic hazard and risk assessment as required by the ATEX directives.

Explosion containment is a protection measure that involves the use of either "explosion-pressure-resistant" or "explosion-pressure-shock-resistant" equipment designed to withstand, without rupture, the maximum explosion pressure likely to be encountered.

Explosion-pressure-resistant equipment is designed to withstand the explosion pressure without becoming permanently deformed.

Explosion-pressure-shock-resistant equipment is designed to withstand the expected explosion pressure, but may become permanently deformed.

Some equipment, as hammer mills and certain other grinding equipment, are often strong enough to contain an explosion (ductwork leading to and from them may need supplementary protection). But in general it is not practicable to produce plant capable of withstanding the typical dust explosion overpressures in the range of 8–10 bar, unless the equipment is of small volume and simple circular or spherical shape.

A combination of some equipment reinforcement and the installation of relief vent(s) may in some cases be an appropriate risk minimising measure.

10.7 Fire Risks—Supplementing the Explosion Risks

Some dusts, such as coal, are capable of self heating when they are held in quantity, deposited on heated surface, or deliberately heated as part of the process. Depending on the conditions, the product may rise in temperature until it starts to smoulder or burn. Dusts which demonstrate this hazard can be identified by a variety of tests.

Small smouldering fires may develop in dust accumulations not only from self heating but also from any of the common sources of ignition. If you suspect a fire inside dust handling equipment, it may be very dangerous to open up any of the inspection points. A sudden rush of air into the equipment could cause a smouldering deposit to flare up, or a dust cloud to form, followed by an explosion that vents out through the inspection point.

Precautionary measures usually involve the application of a cooling medium, an inerting medium, or where it can be done safely, emptying (without generating dust clouds) the product from the plant (not recommended for fine coal dust).

If you try to extinguish fire using water, it is important that you apply it as a fine spray or fog. Using high-pressure water jets on a smouldering fire is dangerous, as you can raise dust clouds. Attempts to restrict the spread of fire by removing dust from adjacent plant equipment have also resulted in the unintentional formation of dust clouds with disastrous consequences.

Note: When tackling **fires involving coal you should not use water** as it may cause a violent reaction or the formation of flammable gases (carbon monoxide may be formed). Dry sand applied cautiously to a small burning heap on the floor from long-handled shovels may be effective, but special proprietary powder fire extinguishers are better.

You may tackle deep-seated fires inside dust handling equipment by applying an inert atmosphere (as carbon dioxide or nitrogen). Although being an effective way of extinguishing fires, it is likely to take a considerable time and large volumes of gas for displacement of all the air from the centre of a large volume of powder and it may take days or even weeks to dissipate the residual heat from a fire in a large silo.

A suitable arrangement for closing (preventing oxygen supply) the equipment may also be used.

10.8 Coal Dust Zone Classification

A coal dust explosion risk is determined based on the following **material explosibility parameters**:

- Coal **type**/group.
- **Particle size** (and its distribution). Sizes above 500 μm do not in general entail any explosion risk.
- **Volatile ratio**. Below 12 %, as valid for anthracite, in general exclude explosion risks. But all other coals (ranked by their volatile content) contain sufficient volatile matter to constitute an explosion risk.
- **Moisture content**. For moisture to display any significant explosion risk eliminating effect, the concentration must be above 5 %. Moisture content above 20 %, will in general exclude explosion risks.
- **Quantity**. The amount of coal dust must be above the LEL (Lower explosive Limit) which for coal in general can be set to 40 g/m^3, and below UEL (Upper explosive Level), which in general can be set to 2 kg/m^3 (but is without practical significance in tackling dust explosion risks).

A concentration of 40 g/m^3 in air means that a glowing 25 W light bulb hardly can be seen through the dust cloud at a 2 m distance. Also, a person cannot breathe in an atmosphere containing dust at such a level.

As mentioned previously, a 1 mm coal dust layer of a density of say 500 kg/m^3 can create an area (volume) up to a height of 5 m containing a dust concentration of 100 g/m^3. In other words, if footprints are visible in coal dust on the floor or coal dust is seen on the walls at a plant, there is enough coal dust at that particular location to propagate an explosion if suspended in the air.

The quantity in the air is also one of the **process explosibility parameters** consisting of:

- **Suspension**. A sufficient premixing of coal dust and air within the explosive limits is required.
- **Oxygen concentration**. Minimum 12 % oxygen concerning lignite and 14 % oxygen for other coals in order to create an explosive coal dust atmosphere.
- **Effective ignition source**. Here it is of special importance to focus on hot surfaces, hot materials in the process stream, electrically and mechanically generated sparks.
- **Confinement**. In order for the fuel-oxygen reaction to become more than a flash fire (or a large fireball), a high level of confinement is required for the heat generation to produce the destructive high pressures characteristic for explosions. The reaction may be a deflagration with flame speed low compared to the speed of sound, or a detonation with speeds equal to the speed of sound. The possibility of increasing the turbulence of the reacting mixture will enhance the violence of the reaction.

Suspension outside equipment, e.g. created by a primary explosion, can be mitigated by good housekeeping (which is what could be called "a must" in areas exposed to explosive dusts).

In order to eliminate ignition by hot surfaces, the temperatures stated in Table 10.10—based on layer ignition temperature (LIT) or cloud ignition temperature (MIE), as appropriate—must be complied with.

Notes:

1. The ignition temperature of a coal dust cloud and of a layer of coal dust decreases as the volatile content increases.
2. An increase in size results in a rapid rise in the ignition temperature requirements.
3. The minimum ignition temperature of a layer of coal dust decreases sharply as the thickness of the deposit is increased.
4. Increasing moisture content above 8 % means a dramatically increase in the ignition energy for an explosion, and at 15 % moisture, about ten times more energy is required.

Table 10.10 Surface max temperatures and coal dust type

Dust type	T_{max} based on LIT (°C)	T_{max} based on MIE (°C)	T_{max}
Anthracite	340 − 75 = 265	2/3 × 700 = 460	265
Pet coke	280 − 75 = 205	2/3 × 690 = 455	205
Bituminous, low volatile	260 − 75 = 185	2/3 × 600 = 396	185
Bituminous, high volatile	250 − 75 = 175	2/3 × 550 = 366	175
Lignite	230 − 75 = 155	2/3 × 440 = 290	155
Wood	315 − 75 = 240	2/3 × 470 = 310	240

5. MIE varies with the oxygen content of the atmosphere, volatile content and the amount of fine dust. Ignition is easier with increasing oxygen or volatile content and with the amount of fine coal dust.

All types of coal dust should be regarded as prone to ignition when exposed to frictional sparks of badly maintained machinery or when they become contaminated with tramp metal.

Mechanically generated frictional sparks represent always an explosion risk when the contact speed is above 10 m/s. Speeds below 1 m/s are regarded as being outside explosion ignition possibilities, while speeds in between may be regarded as possible ignition sources.

Note: The comments above are valid concerning the sparks as such, but the friction may cause dangerous heating of the surface.

For mixtures of coal dust and flammable gas, the critical minimum ignition energy is that which affects the gas.

10.8.1 Some Plant Equipment Considerations

Raw coal stockpile

Typically, raw coal has a particle size distribution outside the explosion range and is stored on an outside—open or partly covered—stockpile. The fire and explosion hazards associated with such stockpiles are usually limited to spontaneous combustion. Embers, smouldering and small fires are not uncommon, and any hot material must be prevented of being loaded into a pulverized fuel system.

Raw coal storage bin

Coal in a bin may be susceptible to spontaneous combustion, although some airflow is required to provide the oxygen necessary for heating. Thermocouples and/or carbon monoxide sensors may be used for detecting an incipient fire.

In the event of an outbreak of fire in the raw coal in a silo, discharge must be possible without sending the coal to the mill.

Risk of explosion—included methane explosion—is considered to be negligible when the top of the bin is appropriately ventilated in order to keep the concentration of methane at a low level.

Mill (Coal pulverizer)

Under normal operating conditions, coal is dropped from a weigh scale into a rotary airlock before it enters the pulverizer. The rotary airlock allows the coal and its inherent moisture to enter the pulverizer, but prevents any outside air from entering the system and acts as an explosion barrier.

As the coal is being ground, hot air normally enters the bottom of the pulverizer and passes up through the pulverizer. The air is used for its drying and conveyance abilities.

The main explosion hazard associated with a pulverizer is related to start-up and shut-down procedures. When a system goes down under load, all the coal falls out of suspension. The internal surfaces are at elevated temperatures and the process of spontaneous combustion begins immediately. If the system is restarted without full knowledge of the internal conditions, an explosion could occur when the hot particles are suspended.

Presupposing normal operation at low concentrations of oxygen (inert conditions), the interior of a pulverizer and associated cyclone filter can be classified as a zone 21 (following the guideline saying, that inerting leads to a one step down in zone classification—here from zone 20 to zone 21). In actual practice, inert conditions (equivalent to a no-zone classification) cannot be guaranteed all the time.

Dedusting filter

The duct leading to a bag filter (or an electric precipitator) should be inclined upwards with an explosion flap on top of the diverter. The venting device will prevent any explosion advancing to the precipitator/filter.

A precipitator can be preceded by a cyclone which should be designed so that the discharge air flow never exceeds 30 g coal dust per m^3 air.

Secondary filters used for the cleaning of coal dust laden air should operate on the "jet-pulse" principle and generally be in an explosion pressure shock resistant design sufficient to withstand the reduced explosion pressure brought about by explosion vents.

As the LEL is not exceeded during foreseeable malfunction with holes in the bags, the internal zone (on the clean side) is regarded as non classified.

System fan

Coal mill filter fans are to be located on the clean air side, i.e. after the dust filter. Internal zone assessed to be zone 22.

Coal meal bin

The storage bin for the pulverized coal will generally contain a potentially explosive coal dust atmosphere, and is to be classified as a zone 20.

Oxidation may take place in the bin.

Depending on the coal type and storage conditions, coal meal may spontaneously ignite in the bottom zones during long duration stops. In order to reduce this hazard, it must be possible to seal off the bottom area before or after the extraction point. The safety is enhanced when bins only containing 4 h coal supply, and they should be foreseen to be emptied before shouting-down.

Carbon dioxide or nitrogen should be foreseen for use in controlling any fire.

10.8.2 Risk Assessment—Inside Equipment

Inside the coal firing system, explosive coal dust atmospheres will exist in varying degrees, covering the zones: 20, 21 and 22, and areas outside zone classification.

As far as possible, appropriate categorized equipment will be installed in the zone classified areas, but the presence of possible effective ignition sources cannot always be excluded. Therefore the possibility of an explosion has to be accepted, and the effect of any explosion must be mitigated.

It is recommended that coal firing systems are designed to operate on a sufficient safe level by the use of explosive-resistant design (explosion-pressure-resistant or explosion-pressure-shock-resistant) and explosion venting systems, alone or in combination. Explosion venting must always entail the release of the exploding atmosphere to safe areas.

10.8.3 Risk Assessment—Outside Equipment

The possibility of an explosive coal dust atmosphere being present outside coal firing equipment is normally very limited, and will never exceed a zone 22 level, provided good housekeeping is adhered to.

At plant layout planning full consideration has to be taken with regard to the need of maintaining a good (or otherwise specified) housekeeping level. Good access points are to be provided to enable regular cleaning of all relevant surfaces, and additional equipment covers or shelters may be relevant.

Where cleaning access is not considered sufficiently feasible, the equipment protection levels may require enhancement.

Close to explosion venting areas unauthorized access must be prevented.

10.8.4 Methane Explosion Aspects

Coal seams and freshly mined coal contains varying amounts of gaseous substances as methane, carbon dioxide, nitrogen and water together with traces of higher hydrocarbons, oxygen, hydrogen and helium.

Coal from surface mining contains only minor amounts of methane, not exceeding 0.1 m^3/ton coal.

Underground coal seams show a very large variation in methane content, generally between 3 and 25 m^3/ton coal.

Freshly mined coal will relatively rapidly lose methane down to the saturation point at 1 atm, which is in the vicinity of 1.2 m^3/ton coal.

Literature data on the subsequent release of methane is very limited.

Methane explosions have occurred in raw coal silos, but it is extremely difficult to say something more precisely about the level of risk—except that the risk seems to be very low for old mined coal and when the air (oxygen) content falls in the inert area.

Measurements in some silos (open with free air flow above) of the methane concentration in the open space above the coal pile gave values below 1 %, and no significant concentration changes was observed throughout the unfilled portion of the silos investigated.

Inside the coal pile, the concentration of methane has been calculated and measured at much higher values (up to 35 and 40 %—which is outside the explosion range). During the reclaiming of the coal, these high methane concentrations could create small areas containing methane explosive atmospheres.

Conclusion: Raw coal silos must have good ventilation, in which case the methane concentration will be outside the explosion range with a reasonable safety margin. Good ventilation is provided by the installed dedusting equipment and room (general) ventilation in case of silos placed inside closed structures.

10.8.5 Equipment Selection

Note: More information regarding equipment selection can be found in Chap. 5.

The key parameters in equipment selection are:

- Equipment category corresponding to relevant zone (or higher),
- Suitability for dusty atmospheres (indicated in Ex-marking by "D"),
- Max surface temperature,
- Protection method, and
- Ingress protection (IP) for electrical equipment surrounded by conductive dust (dust group III),

as long as the explosion risk assessment does not indicate special aspects to be taken in consideration.

Equipment (possessing own ignition source) to be used or installed in zone 20, 21 and 22 must as minimum be categorizes as Ex II 1 D, Ex II 2 D and Ex II 3 D, respectively—unless the explosion risk assessment prescribes otherwise.

(A) **Electrical equipment**

Important standards on electrical equipment (and EX components) for use in potentially explosive dust atmospheres are indicated in Table 10.11.

Table 10.11 Electrical equipment protection type/standard

Standard	Subject/type of protection	ATEX protection code
EN 60079-0:2012	General requirements	–
EN 60079-11:2012	Intrinsic safety	iD
EN 60079-18:2015	Encapsulation	mD
EN 60079-31:2014	Protection by enclosure	tD
EN 60079-2:2015	Protection	pD
EN 60079-14:2014	Design, selection, erection	–
EN 60079-17:2014	Inspection, maintenance	–
EN 60079-19:2011	Repair, overhaul, reclamation	–

The fundamental standard, EN 60079-0 contains specifications that categorize the equipment in a slightly different way, compared to ATEX directive 2014/34, using the Equipment Protection Level (EPL) as a basic parameter as illustrated in Table 10.12.

The different types and levels of protection are to be connected to the required EPL level as stated in Table 10.13.

The combination of the obligatory marking according to ATEX directive 2014/34 and a marking based on the specifications of EN 60079-0—supplemented by temperature indications and Ingress Protection (IP)—can lead to a rather confusing marking on electrical equipment.

The term **Ingress Protection** (IP) is used to identify the environmental protection of an electrical device. The IP code consists of "IP" followed by a two-digit code. The first digit is an indication of the degree of protection against solid objects (as dust) and the second digit informs about the protection against liquid (water). Where the second digit is without significance, it is stated as an "X".

Table 10.12 Equipment specification according to standard and directive

EN 60079-0		ATEX directive 2014/34		
EPL	Group	Group	Category	Zone
Ma	I	I	M1	NA
Mb			M2	
Ga	II	II	1G	0
Gb			2G	1
Gc			3G	2
Da	III		1D	20
Db			2D	21
Dc			3D	22

Table 10.13 Link between EPL and protection

EPL type and level	ATEX symbol	Protection
Da	ta	Enclosure
	ia	Intrinsic safety
	Ma	Encapsulation
Db	tb	Enclosure
	ib	Intrinsic safety
	mb	Encapsulation
	P	Pressurization
Dc	tc	Enclosure
	Mc	Encapsulation
	p	Pressurization

The IP categories IP5X and IP6X are of special relevance concerning dust protection:

Zone (dust)	ATEX category	Required IP
20 (all)	Min 1D	IP6X
21 (all)	Min 2D	IP6X
22 (conductive)	Min 3D	IP6X
22 (non-conductive)	Min 3D	IP5X

IP5X: Dust-protected. Limited dust ingress. No harmful deposit
IP6X: Dust tight. No dust ingress

Marking of ATEX electrical equipment is dealt with in Chaps. 5 and 16.

Ex-marking examples
Ex ma IIIC T 120 °C Da (equipment for use in conductive dust of group IIIC requiring IP6X, with protection m on a EPL Da level with max surface temperature of less than 120 °C).

Ex tb IIIC T 225 °C T_{500} 320 °C Db (equipment for use in conductive dust of group IIIC requiring IP6X, with protection "t" on a EPL Db level with a max surface temperature of less than 225 °C and less than 320 °C when tested with a 500 mm dust layer).

(B) **Non-electrical equipment**

Important standards on non-electrical equipment for use in potentially explosive dust atmosphere are presented in Table 10.14.

A proposal (prEN ISO 80079-36) for modifying EN 13463-1 has been put forward primarily in order to create alignment with the specification of EN 60079-0, as presented in Table 10.14.

Table 10.14 Mechanical equipment protection type/standard

Standard	Subject/type of protection	ATEX protection code
EN 13463-1	Basic methods and requirements	h
EN 13463-2	Flow restricting enclosure	fr
EN 13463-3	Flame-proof enclosure	d
EN 13463-5	Construction safety	c
EN 13463-6	Control of ignition source	b
EN 13463-8	Liquid immersion	k

EN 13463-1 is the basic standard for non-electrical equipment which can be used on its own regarding ATEX protection, and it contains valuable specifications on carrying out "Ignition hazard assessments" and assessments of possible ignition sources.

The **marking of non-electrical equipment** is dealt with in Chaps. 5 and 16.

Ex-marking examples

Ex dh III T 120 °C Db (equipment for use in dust, protected based on Flame-proof enclosure, where category 2 equipment is required, having a max surface temperature of 120 °C).

References

EN 600079-10-2 "Explosive atmospheres - Part 10-2: Classification of area - Combustible dust atmospheres"

EN 1127-1 "Explosion prevention and protection"

Chapter 11
ATEX—Worldwide

11.1 Introduction

This chapter provides a broad overview of hazardous area classifications and the parameters involved in protections techniques and equipment selection—worldwide.

Currently there are two systems in use to classify these hazardous areas: The Class/Division system and the Zone system. The Class/Division system is used predominantly in the US and Canada, whereas the rest of the world generally uses the Zone system.

Furthermore, on the international scene you will meet two different equipment temperature code systems for gas explosive atmospheres, two different ingress protection type classification schemes and different equipment marking systems.

11.2 The Class/Division System

In the Class/Division system hazardous areas are classified according to Class, Division and Group (in US described in the Articles 500 and 505 of the NEC (National Electric Code) code).

11.2.1 Classes

Locations are classified depending on the general nature (or properties) of the materials that may be present:

- **Class I**: A location in which flammable gases or vapours are or may be present in the air in quantities sufficient to produce an explosive atmosphere.

T. Jespen, *ATEX—Explosive Atmospheres*, Springer Series
in Reliability Engineering, DOI 10.1007/978-3-319-31367-2_11

- **Class II**: A location in which combustible dust are or may become present in the air in quantities sufficient to produce an explosive atmosphere.
- **Class III**: A location in which easily ignitable fibres or flyings may become airborne in quantities sufficient to produce ignitable mixtures.

11.2.2 Divisions

The division number is an indication of the probability of the hazardous material to be able to produce an explosive or ignitable mixture:

- **Division 1** indicates that the hazardous material has a high probability of producing an explosive or ignitable mixture due to it being present continuously, intermittently, or periodically or from the equipment itself under normal operating conditions.
- **Division 2** indicates that the hazardous material has a low probability of producing an explosive or ignitable mixture and is present only during abnormal conditions for a short period of time.

11.2.3 Groups

The group defines the type of hazardous material in the surrounding atmosphere. The Groups A, B, C, and D are for gases (i.e. only relevant concerning Class I), while the Groups E, F, and G are for dusts and flyings (relevant for Class II or III):

- **Group A**: Atmospheres containing acetylene.
- **Group B**: Atmospheres containing a flammable gas, flammable liquid-produced vapour, or combustible liquid-produced vapour possessing comparable characteristics with hydrogen, butadiene, ethylene oxide, propylene oxide, and acrolein.
- **Group C**: Atmospheres containing a flammable gas, flammable liquid-produced vapour, or combustible liquid-produced vapour possessing comparable characteristics with ethyl-ether, ethylene, acetaldehyde, and cyclopropane.
- **Group D**: Atmospheres containing a flammable gas, flammable liquid-produced vapour, or combustible liquid-produced vapour possessing comparable characteristics with acetone, ammonia, benzene, butane, ethanol, gasoline, methane, natural gas, naphtha, and propane.
- **Group E**: Atmospheres containing combustible metal dusts such as aluminium, magnesium, and their commercial alloys.
- **Group F**: Atmospheres containing combustible carbonaceous dusts with ≥8 % trapped volatiles, such as carbon black, coal, or coke dust.

- **Group G**: Atmospheres containing combustible dusts not included in Group E or F. Typical dusts include flour, starch, grain, wood, plastic, and chemicals.

11.3 The Zone System

The Zone system is identical with the EU ATEX Zone classification system as described in Chap. 2.

This system has also been adopted in US, as described in Articles 505 and 506 of the NEC code.

11.4 Comparing the Class/Division and the Zone System

Concerning the area classification, Table 11.1 shows the classifications regarding flammable gases and Table 11.2 concerning combustible dust.

The grouping of gases and dusts is not an integrated part of the EU ATEX directives, but the EN standards can be regarded as covering the world outside North America, see Table 11.3.

Table 11.1 Area classification, flammable gases

	Present continuously	Present intermittently	Present abnormally
EU/EN	Zone 0	Zone 1	Zone 2
US: NEC 505	Zone 0	Zone 1	Zone 2
US: NEC 500	Division 1		Division 2

Table 11.2 Area classification, combustible dust

	Present continuously	Present intermittently	Present abnormally
EU/EN	Zone 20	Zone 21	Zone 22
US: NEC 506	Zone 20	Zone 21	Zone 22
US: NEC 500	Division 1		Division 2

Table 11.3 Gas groupings

Typical gas	US (NEC 505) EN	US (NEC 500)
Acetylene	Group IIC	Class I/Group A
Hydrogen	Group IIC	Class I/Group B
Ethylene	Group IIB	Class I/Group C
Propane	Group IIA	Class I/Group D
Methane	Group I (and Group IIA)	Mining[a]

[a]Not within the scope of NEC (under jurisdiction of MSHA)

Table 11.4 Possible future groupings of dust

Dust group	Type	Definition	Examples
IIIA	Combustible flyings	Can burn or glow in air	Rayon, cotton, sisal, jute, hemp, oakum, waste kapok
IIIB	Non-conductive combustible dust	Resistivity $>10^3\ \Omega \times m$	Corn, custard powder, cellulose acetate, nylon, sugar, polythene, epoxy
IIIC	Conductive combustible dust	Resistivity $\leq 10^3\ \Omega \times m$	Aluminium, magnesium, tin, cadmium, coal dust[a]

[a]Not all types of coal are to be regarded as conductive dust

Table 11.3 clearly illustrates the big difference between the North American system and the EN groupings (which represents the European approach).

The unpublished EN 60079-20-2 standard contains a suggestion for the grouping of dusts (solid particles, including fibres, ≤500 μm (0.5 mm)), as indicated in Table 11.4.

This grouping is not in harmony with the grouping (Group E, F and G) under the Class/Division system.

11.5 Explosive Gas Atmosphere Equipment Temperature Codes

The EN, ATEX and US temperature classes are presented in Tables 11.5 and 11.6.

11.6 Ingress Protection Classification Systems

EN 60529 classifies the sealing effectiveness of enclosures of electrical equipment against the intrusion of foreign bodies (dust and moisture) by the IP (Ingress Protection) code: The letters "IP" followed by two digits (An "X" is used for one of the digits if there is only one class of protection).

Table 11.5 EN/ATEX and US, NEC 505 temperature classes

Code	Max equipment surface temperature	
	°C	°F
T1	450	842
T2	300	572
T3	200	392
T4	135	275
T5	100	212
T6	85	185

Table 11.6 US, NEC 500 temperature classes

Code	Max equipment surface temperature	
	°C	°F
T1	450	842
T2	300	572
T2A	280	536
T2B	260	500
T2C	230	446
T2D	215	419
T3	200	392
T3A	180	356
T3B	165	329
T3C	160	320
T4	135	275
T4A	120	248
T5	100	212
T6	85	185

The first digit (0, 1, 2, 3, 4, 5 or 6) indicates the protection against intrusion of foreign objects (dust) and the second digit (0, 1, 2, 3, 4, 5, 6, 7 or 8) indicates the protection against water intrusion, as illustrated in Table 11.7.

In North America, environmental protection is classified using the NEMA (National Electrical Manufacturing Association) standard Publ. No. 250, see Table 11.8.

A non-binding guideline concerning some equivalence between EN and NEMA classifications shown in Tables 11.7 and 11.8, is presented in Table 11.9.

Table 11.7 IP classes

1st Digit	Dust protection	2nd digit	Water protection
0	No protection	0	No protection
1	Dust >50 mm	1	Drops of water
2	Dust >12 mm	2	Vertically dripping water
3	Dust >2.5 mm	3	Sprayed water
4	Dust >1 mm	4	Splashed water
5	Dust-protected	5	Low-pressure water jets
6	Dust-tight	6	Powerful water jets
		7	Effects of immersion
		8	Continuous immersion

Table 11.8 North American environment protection classification

A degree of protection against the following conditions	Enclosure type											
	2	3	3R	3S	4	4X	5	6	6P	12	12K	13
Dripping and light splashing of non-corrosive liquids and falling dirt	X	X	X	X	X	X	X	X	X	X	X	X
Circulating dust, fibres and flyings	–	X	–	X	X	X	–	X	X	X	X	X
Settling airborne dust, fibres and flyings	–	X	–	X	X	X	X	X	X	X	X	X
Hose-down and splashing water	–	–	–	–	X	X	–	X	X	–	–	–
Corrosion	–	–	–	–	–	X	–	–	X	–	–	–
Occasional temporary submersion	–	–	–	–	–	–	–	X	X	–	–	–
Occasional prolonged submersion	–	–	–	–	–	–	–	–	X	–	–	–
Oil and coolant seepage, spraying and splashing	–	–	–	–	–	–	–	–	–	–	–	X
Rain, snow and external formation of ice	–	X	X	X	X	X	–	X	X	–	–	–
External formation of ice	–	–	–	X	–	–	–	–	–	–	–	–
Wind-blown dust	–	X	–	X	X	X	–	X	X	–	–	–

Table 11.9 Best equivalence between EN and NEMA classifications

Degree of protection NEMA	Degree of protection EN
1	IP10
2	IP30
3	IP55
3R	IP55
3S	IP55
4 and 4X	IP66
5	IP52
6	IP67
12	IP55 and IP65
13	IP65

11.7 Electrical Protection Concepts

The electrical protection concepts (Ex code and IEC standards) presented in Table 11.10 are of general validity.

Table 11.10 Electrical EN standards and protection type/code

EN standard	Protection type	Ex code
60079-0	General standard	
60079-1	Flameproof	d
60079-7	Increased safety	e
60079-15	Type n (non-incentive)	n
60079-2	Pressurised/purge	p
60079-11	Intrinsically safe	i
60079-18	Encapsulation	m
60079-6	Oil emission	o
60079-5	Powder filling	q
60079-31	Enclosure (dust)	tD

11.8 Mechanical Protection Concepts

At present, the supporting specifications to be fulfilled by non-electrical equipment are to be found in the EN standards, listed in Table 11.11.

Most Category 3 mechanical equipment needs only to meet the requirements put down by the general standard EN 13463-1.

At the moment (for several years), the EN standards are being used as a basis for developing the international standards (ISO/EN):

- prEN ISO 80079-36 “Non-electrical equipment for use in explosive atmospheres—Basic methods and requirements” will cover the field in line with EN 13463-1.
- prEN ISO 80079-37 will cover the protection types “c”, “b” and “k”, using EN 13463-5, -6 and -8 as the basis.

Protection “fr” (EN 13463-2 on Protection by flow restricting enclosure) and “d” (EN 13463-3 on Protection by flameproof enclosure) are not intended to be included in the ISO/IEC standards. Should these concepts need to be applied internationally, it is foreseen that the corresponding electrical standards can also be used for non-electrical equipment.

Table 11.11 Mechanical EN standards and protection type/code

EN standard	Protection type	Ex code
13463-1	General standard	
13463-2	Flow restricting enclosure	fr
13463-3	Flameproof enclosure	d
13463-5	Constructional safety	c
13463-6	Control of ignition source	b
13463-8	Liquid immersion	k

11.9 Marking of ATEX Equipment

Marking (Ex-marking) of equipment according to ATEX equipment directive 94/9 and EN 60079 are dealt with in Chaps. 5 and 16.

Approved equipment according to the **Class/Division system** is marked with Class (I, II, or III), Division (1 or 2), Group (A, B, C, D, E, F, or G), and the temperature code (T1 through T6) as appropriate. For intrinsically safe equipment the words "Intrinsically Safe" or just "IS" will precede the actual approval marking to indicate it as being intrinsically safe.

Examples:

- Class I Division 1 Group B, C, D T5
- CL I Div 2 GP ABCD T5
- IS CL I, II, III Div 1 GP ABCDEFG
- CL II, III Div 1, 2 GP EFG T4

As North America has accepted the Zone system, the Division specification may be exchanged with a Zone specification, so a Class I, Division 1, Group D product may also be marked as Class I, Zone 0, Group IIA.

But it is necessary to be very careful when division-based equipment is being zone-classified (For example, not all Division 1 equipment is suitable for Zone 0).

11.10 Approval Agencies

Most countries require that equipment intended for installation in explosive atmosphere hazardous areas has been approved (at least concerning equipment for the more hazardous areas) by a recognized authority or approval agency (as the Notified Bodies in EU), which that country has established by law, regulation or code.

In US some of the 15 national testing laboratories (NRTL's) are qualified to approve products for use in hazardous areas.

Countries participating in the IECEx scheme (International Electrotechnical Commission on explosion protected equipment, known as "Ex") can issue either an international certification or a national certification of explosion protected equipment.

Each country within the IECEx scheme establishes an ExCB (Ex Certification Body) which can approve products. ExCB's can issue the national certification for their country based upon the IECEx standards (including any national deviations) and the international certification. Currently, Australia is the only country accepting international certifications for use in their country.

Part II
ATEX Equipment Safety Aspects—Manufacturer Obligations

Chapter 12
ATEX—Equipment Legislation

12.1 Introduction

The objective of this chapter is to continue the introduction to the ATEX field, as presented in Chap. 2 and to set focus on the ATEX Equipment directive 2014/34 in order to provide an overview of the obligations placed on the Economic Operators. Among them, especially the manufacturer of equipment and protective systems intended for use in hazardous areas due to potentially explosive atmospheres (Zone classified areas) will be in focus.

A more comprehensive introduction to the ATEX equipment directive 2014/34 is the European Commission's ATEX Guidelines (4th edition—September 2012) that is primarily intended to facilitate the correct application of the directive.

Directive 2014/34 replaced the previous ATEX equipment directive 94/9 in April 2016.

Key topics as:

- The Essential Health and Safety Requirements specified in Annex II to the Directive,
- The Conformity assessment procedures, and
- The CE- and Ex-marking.

have not been changed, which means that the replacement does not affect the relevant ATEX EN standards.

The major changes concern provisions on third-part (Notified Bodies) conformity assessments tasks and market surveillance control, which are subjects outside the scope of this book.

Some minor changes (but certainly not of minor importance) are the following:

- Distinct listing of the obligations of the Economic Operators: Manufacturers, Authorised representatives, Importers and Distributors,
- Multiple new definitions,

T. Jespen, *ATEX—Explosive Atmospheres*, Springer Series
in Reliability Engineering, DOI 10.1007/978-3-319-31367-2_12

- Emphasis on ensuring traceability of a product throughout the whole supply chain in order to make market surveillance simpler and more efficient,
- The "EC Declaration of conformity" changes name to "EU Declaration of Conformity",
- ATEX equipment only to have one conformity declaration (the EU Declaration of Conformity) that covers all the requirements of the directive and other applicable directives, and
- The "EC-Type Examination Certificate" changes name to "EU-Type Examination Certificated".

12.2 Scope of the ATEX Equipment Directive 2014/34

ATEX directive 2014/34 applies to equipment and protective systems intended for use in potentially explosive atmospheres and to devices (safety devices, controlling devices and regulating devices) outside such areas influencing the safe functioning of mentioned equipment and protective systems.

Note: Only equipment capable of causing an explosion through its own potential source of ignition is included (This criteria is not valid for protective systems and mentioned devices).

Protective systems may be integrated into equipment or separately placed on the market for use as autonomous systems, as flame arresters, relief vent systems, suppression systems or extinguish barriers. Only protective systems separately placed on the market fall under the direct regulation of the directive of such systems.

Under the listing (definition) of "equipment" you also finds "components" defined as any item essential to the safe operation of equipment and protective systems but with no autonomous function.

Concerning the definition of a "potential explosive atmosphere", see Chap. 2.

12.3 Assemblies

An assembly is also a product covered by the directive when it is formed by combining two or more pieces of equipment, together with components if necessary, and that assembly is placed on the market (or put into service) as a single functional unit. Such assemblies will often require proper installation at the place of use.

When a final combination of equipment (e.g. some ATEX mechanical equipment which needs to be connected to different pieces of ATEX electrical equipment) does not create additional ignition hazards (determined on the basis of an Ignition hazard assessment (IHA)—see subsequent Chap. 13), no further action is required. The compliance of the assembly is then based on the compliance of the individual parts.

The assembly manufacturer may also regard the assembly as a piece of ATEX equipment, affix CE- and Ex-marking, sign a EU Declaration of conformity, and provide instructions for the safe use of the assembly. In doing so, the assembly manufacturer takes complete responsibility, and this procedure does not require any renewed involvement of a Notified Body.

Where an assembly consists of parts not already in full conformity with relevant parts of the directive, or the Ignition hazard assessment has revealed additional/new risks, the assembly must undergo the complete conformity assessment procedure appropriate for the category.

12.4 Equipment Groups and Categories

Annex I of the directive lays down criteria determining the classification of equipment groups and the categorization of equipment according to the level of protection ensured by the equipment.

The philosophy is then to require (following ATEX Workplace directive 1999/92) the use of equipment based on a high level of protection in areas where the occurrence of an explosive atmosphere has a high probability and vice versa.

The directive divides equipment into two groups:

- **Group I** comprises equipment for use in mining areas likely to become endangered by firedamp and/or combustible dust.
- **Group II** comprises equipment for use on other places likely to become endangered by explosive atmospheres.

These groups are subdivided into Categories as follows.

Group I is divided into:

- **Category M1** comprising equipment which can remain functional when an explosive atmosphere is present, and is characterized by special integrated protection measures ensuring a very high level of protection.
- **Category M2** comprises equipment designed to be capable of remaining within its operational parameters and ensuring a high level of protection (and intended to be de-energised in the event of an explosive atmosphere).

Group II is divided into:

- **Category 1** comprises equipment capable of functioning in conformity with its operational parameters and ensuring a **very high level** of protection (intended for use in Zone 0 and Zone 20).
- **Category 2** comprises equipment capable of functioning in conformity with its operational parameters and ensuring a **high level** of protection (intended for use in Zone 1 and Zone 21).

- **Category 3** comprises equipment capable of functioning in conformity with its operational parameters and ensuring a **normal level** of protection (intended for use in Zone 2 and Zone 22).

The intended use, as determined by the manufacturer will place the equipment in question in the relevant group/category combination, which again determines the appropriate conformity assessment procedure to be followed, including the essential health and safety requirements (EHSR) to be fulfilled (specified in Annex II of the directive).

Note: Contrary to the Machinery directive 2006/42 you cannot just declare compliance with the ATEX equipment directive 2014/34 without specifying equipment type (e.g. electrical or mechanical) and relevant Group and Category.

Note: The use of Group I, Category M1 and M2 equipment are not regulated by ATEX Workplace directive 1999/92 (which defines the zoning of hazardous areas due to the possible occurrence of explosive atmospheres).

Note: Protective systems placed on the market for use as autonomous systems are not characterised by Group and Category.

12.5 Obligations of Economic Operators

The obligations of the Economic Operators (Manufacturers, Authorised representative, Importers and Distributors) are stated in this chapter of the directive.

The following subsections contain some of the more important (in the author's opinion) requirements. If you belong to one of the operator categories, you should of course acquaint yourself with all the provisions belonging to your category.

12.5.1 Manufacturer

A manufacturer is a natural or legal person who manufactures a product (**or** let others design or manufacturer it) **and** market it under his name or trade mark **or** uses it himself.

Manufacturer obligations:

- Ensure compliance with EHSR of Annex II when the product is placed on the market or put into service by himself.
- Draw up the Technical documentation (See Sect. 12.9) and carry out the appropriate Conformity assessment procedure (See Sect. 12.7).
- For a product other than a component: Draw up an EU Declaration of Conformity (See Sect. 12.8) and affix the CE-marking (See Sect. 12.11).
- For a component: Draw up a written Attestation of Conformity (See Sect. 12.8).
- Ensure (in general) that the product is accompanied by a copy of the Declaration or the Attestation, as appropriate.

- Ensure that the products placed on the market bear (in general) a type, batch or serial number or other element allowing their identification.
- Ensure that products other than components, placed on the market, bear the specific explosion protection marking (See Sect. 12.11).
- Indicate on the product (could be on the packaging) their name, trade name or mark and the postal address, easy to understand and use for contact.
- Ensure that the product is accompanied by Instructions and safety information (See Sect. 12.10), easily understood by end-users.
- Based on a reasoned request from a competent national authority, provide all the information necessary to demonstrate compliance with the directive to that authority in a language easily understood by that authority.

Note: The person indicated as the manufacturer on the product is responsible for compliance with the directive, including Declaration and CE-marking, when the product is placed on the market or put into service in EU. The person may be established inside or outside EU, and does not have to be the person who in reality manufactured the product.

Note: "Placing on the market" means the first making available of a product on the EU market.

12.5.2 Authorised Representative

An Authorised representative is a natural or legal person established within EU who by a written mandate from a manufacturer is authorised on his behalf to carry out specific tasks.

Authorised representative obligations:

- Perform the tasks specified in the mandate, which must at least include the following: (1) Keep the EU Declaration or the Attestation of Conformity, as appropriate, at the disposal of the surveillance authorities for 10 years, (2) Provide that authority with the necessary information to demonstrate conformity—after having received a reasoned request. (3) Cooperate with competent national authorities where appropriate to eliminate risks posed by the product(s).
- His obligations, specified in the mandate **must not** include manufacturers obligations to comply with the EHSR of Annex II, to draw up the Technical documentation and to carry out the appropriate Conformity assessment procedure.

Note: The manufacturer issuing the mandate may be established inside or outside EU.

12.5.3 Importer

An importer is a natural or legal person established within EU who places a product from outside EU on the EU market.

Importer obligations:

- Before placing a product on the EU market, ensure that the manufacturer has carried out the appropriate Conformity assessment procedure, that he has drawn up the Technical documentation, that the product bears correct and relevant markings (including manufacturer identification), and that the product is accompanied by the EU Declaration or Attestation of conformity, as appropriate.
- Indicate on the product (in some cases on the packaging or in an accompanying document) the name, trade name or trade mark and the postal address of the importer.
- Ensure that the product is accompanied by instructions and safety information in a language easily understood by end-users.
- Keep for 10 years a copy of the Declaration or the Attestation at the disposal of the relevant authorities, and ensure that the Technical documentation can be made available to those authorities upon request.
- Based on a reasoned request from a competent national authority, provide all the necessary information to demonstrate compliance with the directive to that authority in a language easily understood by that authority.

12.5.4 Distributer

A distributer is a natural or legal person in the supply chain, other than the manufacturer or the importer, who makes a product available on the market.

Distributor obligations:

- Verify that the product bears correct and relevant markings (including manufacturer and importer identification, as appropriate), is accompanied by Declaration or Attestation, as appropriate, by instructions and safety information in a language easily understood by end-users.
- Based on a reasoned request from a competent national authority, provide all the necessary information to demonstrate compliance with the directive.
- Act with due care in relation to the requirements of the directive (and behave as a responsible and conscientious person).

12.6 Essential Health and Safety Requirements (EHSR)

Equipment, protective systems, and devices must meet the EHSR set out in Annex II of the directive according to their intended uses.

The EHSR includes general requirements (Annex II, point 1) concerning for example markings, instructions, selection of materials, design and construction, potential ignition sources, hazards arising from external effects, and safety related devices.

Annex II, point 2 contains supplementary requirements for equipment according to Group and Category. The Category (1, 2, and 3) specific requirements for equipment in Group II, distinguish between intended use in explosive atmospheres caused by gasses, vapours or mist and explosive atmospheres caused by dust.

Annex II, point 3 contains supplementary requirements for protective systems.

In order to comply with the stated EHSR the manufacturer is supported by several applicable standards in the series EN 60079 (covering electrical equipment) and EN 13463 (covering mechanical equipment).

12.7 Conformity Assessment Procedures

The possible procedures for assessing the conformity of equipment and autonomous protective systems (and where necessary devices) are described in Article 13 of the directive with references to the Annexes III, IV, V, VI, VII, VIII, and IX. The annexes contain assessment procedures that cover both product evaluation and product production stages, and procedures that have to be combined (See below).

- Annex III: EU-Type examination (module B): Of relevance on the product evaluation stage involving a Notified Body who issues an **EU-Type Examination Certificate**.
- Annex IV: Conformity to type based on quality assurance of the production process (module D): Of relevance on the production control stage involving a Notified Body who issues a **Quality Assurance Certificate**, and whose identification shall appear in the marking of the product. An authorised representative may take care of some of the manufacturer obligations.
- Annex V: Conformity to type based on product verification (module E): Of relevance on the production control stage involving a Notified Body who issues a **Certificate of Conformity**, and whose identification shall appear in the marking of the product. An authorised representative may take care of some of the manufacturer obligations.
- Annex VI: Conformity to type based on internal production control plus supervised product testing (module C1): Of relevance on the production control stage involving a Notified Body whose identification shall appear in the marking of the product. An Authorised representative may take care of some of the manufacturer obligations.

- Annex VII: Conformity to type based on product quality assurance (module E): Of relevance on the production control stage involving a Notified Body who issues a **Quality Assurance Notification**, and whose identification shall appear in the marking of the product. An Authorised representative may take care of some of the manufacturer obligations.
- Annex VIII: Internal production control (module A): Covers product evaluation and production control stages **without** involvement of a Notified Body. An authorised representative may take care of some of the manufacturer obligations.
- Annex IX: Conformity based on unit verification (module G): Covers product evaluation and production control stages involving a Notified Body who issues a **Certificate of Conformity**, and whose identification shall appear in the marking on the product. An Authorised representative may take care of some of the manufacturer obligations.

Obligatory Conformity Assessment Procedures

Products belonging to Group I, Category M1 or Group II, Category 1 (relevant for Zone 0 or Zone 20):

- Annex III procedure in conjunction with either the Annex IV procedure or the Annex V procedure.

Products belonging to Group I, Category M2 or Group II, Category 2 (relevant for Zone 1 or Zone 21):

- Internal combustion engines and electrical equipment have to follow the Annex III procedure in conjunction with either the Annex VI procedure or the Annex VII procedure.
- Other products in these Groups/Categories may follow the Annex VIII procedure provided the Technical documentation is communicated to a Notified Body who shall acknowledge receipt and retain the documentation.

Products belonging to Group II, Category 3 (relevant for Zone 2 or Zone 22):

- The Annex VIII procedure.

Products belonging to Group I or Group II:

- In addition to abovementioned procedures, the Annex IX procedure may also be followed.

Conformity assessment regarding **Components** shall follow abovementioned scheme with the exception of affixing the CE-marking and the drawing up of the EU Declaration of Conformity (instead an Attestation of Conformity must be issued).

Note: Only equipment Group II, Category 3 can be assessed without any involvement of a Notified Body. Concerning products other than electrical equipment and internal combustion engines of Group I, Category M2 and Group II, Category 2, you have to involve a Notified Body, but it is sufficient to send the technical documentation to the Notified Body, which shall acknowledge receipt of it and retain it.

Note: Although Group II, Category 3 equipment can be self-certified by the manufacturer, many purchasers may still prefer a third-party certification. Issue of an EU-Type Examination Certificate or Quality Assurance Notification is not permitted for Category 3 equipment.

Note: An authorised representative (appointed by the manufacturer by a written mandate) may act on behalf of the manufacturer in respect of some of the conformity assessment procedure obligations (which can be of special relevance where the manufacturer is not established within the EU).

12.8 Documents of Conformity

Equipment, protective systems and devices must be accompanied by an **EU Declaration of conformity** (and bearing the CE-marking affixed) with reference to the ATEX Equipment directive 2014/34.

Components intended for incorporation into equipment or protective systems must be accompanied by an **Attestation of conformity** (and must not have the CE-marking affixed) with reference to the ATEX Equipment directive 2014/34.

The manufacturer or his Authorised representative in EU shall draw up the Declaration of conformity or the Attestation of conformity, as appropriate.

The **Conformity declaration** shall state that the fulfilment of the essential health and safety requirements has been demonstrated, and shall have the model structure set out in Annex X to the directive. It shall contain the elements specified in the relevant conformity assessment procedures, and be translated into the language(s) required by the Member State in which the product is placed or made available on the market.

In case the product is subject to more than one directive/regulation requiring an EU Declaration of Conformity, only one declaration shall be drawn up, which informs about the different EU legislation covered.

Note: The Declaration shall contain the name and address of the manufacturer and his authorised representative, where applicable.

Note: As compliance with ATEX directive 2014/34 is not an unambiguous declaration, it is of key importance that the declaration clearly specifies which requirements are fulfilled. This specification may be met by stating the relevant Group and Category and Ex-marking.

The **Conformity attestation** issued by the manufacturer, shall declare conformity with the applicable provisions of the directive and stating their characteristics and how they must be incorporated into equipment or protective systems in order to assist in the overall compliance with the relevant EHSR of Annex II to the Directive.

The directive is not very specific on the content of a conformity attestation, but besides a clear identification of the manufacturer, the following must be included in the attestation:

- Information on the characteristics of the components, and give
- Incorporation information (for it to assist compliance with the essential requirements applicable to the finished equipment or protective system).

The characteristics of a component will normally include the assigning to the component an explosion classification in the same way as for ATEX equipment, for instance as: II 1/2 c GD Tx (equivalent to an Ex-marking which is not mandatory for ATEX components).

Note: Although zone classification is used for selecting the appropriate equipment, zone classification criteria are—in principle—not an EU harmonised subject. Zone classification specifications are therefore not an issue for ATEX conformity declarations (and Ex-marking).

Documents issued by a Notified Body, such as EU-Type examination certificates and notifications, need not accompany the product.

12.9 Technical Documentation (Dossier)

Technical documentation requirements are listed in Annex III (EU-Type examination), Annex VIII (Internal control of production) and Annex IX (Conformity based on unit verification) of the ATEX directive 2014/34 as:

- A general description of the equipment.
- The design and manufacturing drawings and layouts/schemes supplemented by explanations—also for understanding the operation of the equipment.
- A list of harmonised standards applied in full or in part supplemented by descriptions of the satisfactory solutions adopted where standards have not been applied.
- Results of calculations, examinations and test reports.

It seems logical to regard certificates and other documents issued by a Notified Body as part of the Technical documentation (which must include all the relevant documentation enabling the conformity of the product with the requirements of the directive to be assessed).

The technical documentation must be compiled in one or more official EU languages and be kept available for at least 10 years by the manufacturer. The same obligation is also valid regarding his Authorised representative, if applicable.

An importer shall ensure that the manufacturer has drawn up the technical documentation and ensure that the documentation can be made available to the authorities, on request.

A similar obligation is placed on a distributor, but here the requirement is not directly linked to "technical documentation", but to documentation demonstrating the conformity of the product.

12.10 Instructions

All ATEX equipment and protective systems must be accompanied by instructions (and safety information), that meet the requirements put down in Annex II, point 1.0.6 of the ATEX equipment directive 2014/34.

The instructions and safety information must be drawn up in a language, which is easily understood by end-users, as determined by the Member State concerned.

12.11 Markings

As part of the conformity assessment procedure, ATEX Equipment directive 2014/34 provides for the affixing of the CE-marking of equipment and protective systems. The CE-marking shall be followed by the identification number of the Notified Body, where such body is involved in the production control stage (See Sect. 12.7).

Equipment and protective systems should not have the identification number of a Notified Body affixed if such products belong to Group II, Category 3 (without having a Certification of Conformity according to the Annex IX procedure), as well as some mechanical Group I, Category M2 and Group II, Category 2 equipment, and any use of voluntary certification.

Furthermore, the marking must include:

- Name, registered trade name or registered trade mark, and address of the manufacturer.
- Designation of series or type.
- Batch or serial number, if any.
- Year of construction.
- Specific explosion protection marking (Ex-marking symbol) followed by the symbol of the equipment Group (I or II) and Category (For Group I: M1 or M2, and for Group II: 1, 2 or 3)
- For equipment Group II, the letter "G" when the intended use is in a gas, vapor or mist atmospheres or the letter "D" when the intended use is in a dust atmosphere (or "D G" when the intended use covers both "D" and "G").
- And **where necessary**, the marking must also contain all information essential to the safe use of the product.

The supplementary necessary information is in general covered by the specification of:

- Ex Code of the used protection type.
- Ingress protection (IP code), as appropriate.
- Max surface temperature.

In practice, the task of marking ATEX equipment and protective systems and attempts to understanding the contents of a marking may be rather complicated. This may be because more than one category can be relevant for a piece of equipment, and because the fundamental ATEX standard EN 60079-0 contains specifications that categorize equipment in a slightly different way compared to the ATEX directive 2014/34—See Chap. 16 for more information.

In any case, **the instructions must explain in detail the meaning of the marking on the product**.

Safety devices, **controlling devices and regulating devices** (**Article 1.1 (b) devices**) intended for use outside ATEX zones, but required for or contributing to the safe functioning of equipment and protective systems with respect to explosion risks, also fall under the scope of the ATEX Equipment directive 2014/34.

Such Article 1.1 (b) devices, separately put on the market are to be CE- and Ex-marked, indicating the Category of equipment to be under their control in round brackets. For example "II (1) G D" indicates a device to be places outside ATEX zones, which can be connected to Group II, Category 1 equipment—for both gas and dust atmospheres. And "II 2(1) G D" indicates Category 2 equipment containing an Article 1.1 (b) device for a Group II, Category 1 equipment—for both gas and dust atmospheres).

Components are not to be CE-marked according to the directive, but recommendations for the opposite point of view are also encountered in the literature.

12.12 Some Important Basic Aspects

Equipment intended for use in potentially explosive atmospheres, is only covered by the ATEX equipment directive 2014/34 when it has its own ignition source. When it is evident that the equipment has no effective ignition source of its own, the product is therefore outside the scope of the directive, **but** in many cases an Ignition hazard assessment (IHA)—see Chap. 13—should be carried out in order to clarify and document this fact.

Protective systems are subject to the directive regardless as to whether they have their own potential ignition source or not.

Electrical equipment intended for use in potentially explosive atmosphere is always regarded as equipment possessing its own possible ignition source. Therefore electrical equipment is not candidates for Ignition hazard assessments.

For electrical equipment you start with deciding the intended use (Group and Category). Next you choose an appropriate protection type and follow the specifications laid down in the matching standard.

Mechanical equipment starts with an Ignition hazard assessment (IHA), where the required degree of detail will depend on the intended use of the equipment (Group and Category).

Of course you can always perform an IHA as detailed as you like, and let the result determined which equipment category that would be a correct classification, which again will determine the specification of the intended use, when the product is placed on the market.

It is of special importance to clarify (and describe) the boundary surface (interface) between the equipment and the surroundings at the intended use of the equipment. The characteristics of the surroundings—determined by the possible existence of an explosive atmosphere or a zone classification—decide the necessary category of the equipment's interface towards each of these individual surroundings.

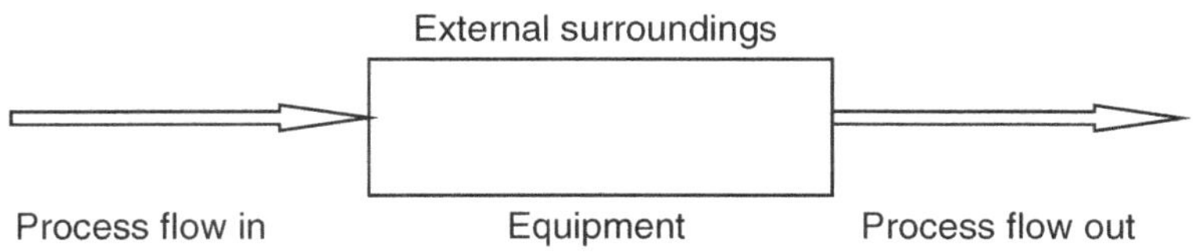

The equipment illustrated above has three interfaces to the surroundings:

- An interface to the external surroundings, e.g. a zone 22
- An interface to the process flow entering the equipment, e.g. a zone 21
- An interface to the process flow leaving the equipment, e.g. a zone 20

The category of the interface of the equipment towards each of these three surroundings must at least fulfil the requirements for the relevant zone classification.

In the example above, it is evident that the equipment must contain an internal explosive atmosphere. However this is not of direct importance regarding the ATEX classification of the equipment as such, **but** this fact must of course be taken into consideration in the general risk assessment of the equipment, and becomes decisive concerning the required category of equipment intended to be placed inside the illustrated equipment.

ATEX equipment classification criteria

The classification of ATEX equipment into three categories is based on the following criteria.

Where the Ignition hazard assessment points to the fact that the product does not contain any effective ignition sources of its own (or they have been avoided) during **normal operation**, the product can be classified as Category 3 equipment.

Where the Ignition hazard assessment moreover points to the fact that the product does not contain any effective ignition sources of its own (or they have been avoided) during **expected malfunctions**, the product can be classified as Category 2 equipment.

Where the Ignition hazard assessment, moreover points to the fact that the product does not contain any effective ignition sources of its own (or they have been avoided) during **rare malfunctions**, the product can be classified as Category 1 equipment.

Therefore, when a classification as Group II, Category 2 equipment is sufficient for the intended use, the assessment may be limited to normal operation and expected malfunctions. And where a Group II, Category 3 classification is sufficient, the assessment may be limited to normal operation—but the inclusion of expected malfunctions is recommended.

Note: Pleading Category 1 or 2 compliance requires the involvement of a Notified Body—at least as a Technical file keeping institution (regarding Category 2 mechanical equipment).

Where the absence of any own possible ignition source is substantiated, the equipment is (legally and on principle) not within the scope of the ATEX Equipment directive 2014/34, which may be tackled and communicated to a customer as described in Chap. 15.

Extended Ignition Hazard Assessment

In general, an Ignition hazard assessment should not only focus on an ATEX equipment categorisation, but be used as an element for being in compliance with the principles of "Integrated explosion safety" and "Inherently safe machinery design and construction". Therefore, it is recommended that everything (alone or in combination) that may trigger an explosion for the equipment in question, is included in the first step(s) of the assessment and not only own ignition sources.

In practice, the most serious possible effective ignition sources may very well not belong to the equipment as such, but are due to elements of the process flow.

An Ignition hazard assessment inspired by EN 13463-1, is described in Chap. 13.

12.13 Standards and Protection Methods

Section 5.5 of Chap. 5 contains an overview of protection types, the standards with the detailed specifications and their relevance according to the intended use, specified by Zone classifications.

The overview covers both electrical and non-electrical (mechanical) equipment, and with one exception regarding EN 13461-1 (see subsequent Sect. 12.14 and Chap. 13) no further details will be given in this publication.

12.14 Basic Requirements and Ignition Hazard Assessment (IHA)

The basic specifications (ensuring compliance with the ATEX equipment directive 2014/34) for mechanical equipment are described in EN 13463-1, which is harmonised to the directive.

Regarding Ignition hazard assessment methods, EN 15198 (2007), which is also harmonised to the directive, contains the presentation of the same ignition hazard assessment method, but in EN 13463-1 you find a more detailed description and illustration of the method. Therefore EN 13463-1 forms the basis for the short presentation of this assessment (in subsequent Chap. 13), which is the key aspect regarding mechanical equipment for use in potentially explosive atmospheres.

As mentioned above, electrical equipment in general contains effective ignition sources, which means that the equipment has to make use of an appropriate explosion protection method before being placed on the market for use in potential explosive atmospheres.

On the other hand, mechanical equipment in normal operation within its design parameters will not trigger the ignition of an explosive atmosphere when properly maintained (and with malfunctions appropriately avoided).

Therefore, a lot of mechanical equipment can comply with Category 3 and even Category 2 equipment requirements by a proper choice of well-tried constructional measures. **But** the determination of the correct category (for the intended use) should be based on an evaluation (and documentation) of the potential ignition sources by the use of a systematic Ignition hazard assessment method.

EN 13463-1 goes through the different possible ignition sources which are described in details in EN 1127-1 (see subsequent Chap. 14), underlines important assessment aspects and specifies the relevant requirements for the different groups and categories of mechanical equipment.

EN 13463-1 also describes "Verification and tests", "Documentation and information for use" and in an informative Annex B "The Ignition hazard assessment procedure" which is illustrated by examples in its Annex C.

Chapter 13
ATEX—Ignition Hazard Assessment (IHA)

13.1 Introduction

While electrical equipment (normally always regarded as possessing some kind of an ignition source) is generally candidates for the application of protective methods when intended for use in potentially explosive atmospheres, mechanical equipment is generally without own ignition sources. Therefore mechanical equipment should always—as the first step—be subject to an Ignition Hazard Assessment (IHA) in order to reveal the existence of any unnoticed effective ignition sources and where disclosed, for determining the eventual equipment Category—taking any implemented protective method into consideration.

The following presentation of an Ignition Hazard Assessment (IHA) is inspired by a model presented in the informative Annex B of EN 13463-1. But the assessment table presented here is more simple and contains fewer columns (8 instead of 16).

13.2 Ignition Source Terms/Definitions

EN 13463-1 defines the following ignition source terms:

- **Possible ignition source** means any kind of ignition source (as listed in EN 1127-1 and EN 13463-1). The listed possible ignition sources form the framework for what may be considered in the IHA.
- **Equipment related ignition source** means any possible ignition source which can be ascribed to the equipment under consideration regardless of its ignition capability. All equipment related ignition sources must be evaluated in the IHA in order to determine whether they are a potential ignition source.

T. Jespen, *ATEX—Explosive Atmospheres*, Springer Series in Reliability Engineering, DOI 10.1007/978-3-319-31367-2_13

- **Potential ignition source** means an equipment related ignition source (own ignition source) which may ignite an explosive atmosphere (The probability of it becoming effective determines the equipment category).
- **Effective ignition source** means a potential ignition source able to ignite an explosive atmosphere taking the probability of occurrence into consideration (in normal operation, expected malfunction or rare malfunction).

The first step of an IHA consists of going through the list (Table 1) of possible ignition sources (12 in all). In this step it is recommended to consider both possible ignition sources that can be ascribed to the equipment as well as possible ignition sources caused by something in the process flow—when the presence of such sources is expected by the manufacturer in relation to the intended use of the equipment. And for equipment related ignition sources, it should be noted if the source is of relevance inside the equipment and/or for the surroundings (the outside) of the equipment.

The subsequent IHA will then set focus on the equipment related ignition sources, but other recognized possible ignition sources should not be forgotten—for example in the risk assessment of the equipment as such.

13.3 Hot Surfaces

"Hot surface" is one of the two very important ignition sources regarding non-electrical equipment (the other is mechanical sparks). Therefore the max surface temperature of the equipment shall be established as part of the IHA.

The max surface temperature, used for the Ex-marking of the equipment, is the highest temperature that can be attained in service under the most adverse operating conditions (but within the recognised tolerance) by any part of the equipment, which can produce an ignition of the surrounding atmosphere. The temperature must be stated with an appropriate safety margin (see clause 8.2 of EN 13463-1 for more information).

The max surface temperature must be established, primarily through appropriate measurements or if not possible, by calculation. The max surface temperature shall in principle reflect a condition only dependent on the equipment itself when operating in the standard temperature range (−20 to +60 °C).

Group IIG equipment (equipment intended for use in explosive atmospheres caused by gases, vapours or mists) may be classified according to the actual max temperature, by the use of temperature class or by stating the specific gas, vapour or mist for which it is intended. Where the max temperature depends not on the equipment itself, but mainly on the operating conditions, the equipment shall be marked with **TX** (and the instruction for use must contain appropriate explanation), indicating that the temperature cannot be marked by the manufacturer.

Group IID equipment (equipment intended for use in explosive atmospheres caused by dust) must be classified according to the actual max temperature. Where the operating conditions are the decisive factor, the **TX** classification is used as for Group IIG equipment.

Chapters 15 and 16 contain more information on the subject.

13.4 Mechanically Generated Sparks

Chapter 14 contains an introduction to this possible ignition source.

The causes behind frictional heating may also produce sparks, and where the generated sparks do not possess sufficient energy to become a possible ignition source (as slow moving parts having a periphery speed of less than 1 m/s), the frictional heating may become a possible ignition source.

Glancing impacts between materials may produce sparks that can cause ignition of flammable atmospheres. While flammable vapour atmospheres may be ignited by a range of this type of spark, usually only sparks involving the thermite reaction caused by impacts between the light metals (aluminium, magnesium, titanium), or their light alloys, and rusty steel present a significant ignition risk to dust clouds.

Control of this potential source of ignition is achieved simply by avoiding the use of the light metals or their light alloys where rust or flammable atmospheres may be present, or placing items made from these metals where they cannot be involved in impacts.

Friction sparks caused by prolonged friction between two materials are more incendive than sparks from a single impact.

Ingress of foreign materials as stones and tramp metals may constitute a bigger ignition hazard than equipment own ignition sources.

13.5 Ignition Hazard Assessment (IHA)—Step by Step

This assessment guideline aims at providing the documentation which fulfils the most essential requirements on the Technical documentation demanded by ATEX Equipment directive 2014/34.

The scheme has two tables:

- Table 13.1 listing the 12 basic possible ignition sources, and
- Table 13.2 to be used for the assessment of the identified equipment related ignition sources.

The IHA may be divided into 4 steps (Sects. 13.5.1–13.5.4):

Table 13.1 Ignition source identification

Possible ignition source	Equipment related	Comments
Hot surface		
Flame and hot gases (including hot particles)		
Mechanical sparks		
Electrical sparks		
Static electricity		
Chemical reactions (including self ignition)		
Lighting		
Stray electric currents and cathodic corrosion protection		
Radio frequency waves		
Optical radiation (including laser radiation)		
Ionizing radiation		
Adiabatic compression and shock waves		

13.5.1 Identification of Relevant Possible Ignition Sources

The first step in the IHA process consists of identifying the relevant possible ignition sources based on the listing in Table 13.1.

During this step it is important to proceed systematically, ignoring any assessment aspects to avoid restriction in thinking. You should not take into consideration specific protective measures, although they are foreseen to be included with the equipment.

In the second column of Table 13.1 you should indicate if the source is equipment related (by “Yes”) or not (by “No”) together with an indication of the source being “inside” and/or “outside” the equipment.

At this stage it is recommended not to restrict the identification of relevant possible ignition sources to those related to the equipment. Therefore, when a “No” is written in column 2, but the possible ignition source may be triggered by the process conditions, this fact should be commented on in the “Comments” column, so it can be carried on to the general risk assessment of the equipment.

Where a “No” covers both the equipment and external circumstances, a “Not present” should be written in column 3 “Comments”.

Note: One type of ignition source may have several causes.

As previously mentioned, ignition sources due to process conditions do not contribute to the ATEX classification of the equipment, but must be taken into consideration, instance in relation to a Machinery risk assessment. In any case, where the manufacturer becomes aware of ignition risks that originate from the process, which cannot be controlled by the design, such information should be conveyed by the "information for use" to the user, drawing attention to the need for implementing additional precautions.

The "Comments" column of Table 13.1 can be used to give more information e.g. about what lies behind the identification of the possible ignition source as being of relevance (what is the reason for its identification)—see example in Chap. 14.

The identified Equipment related ignition sources are then transferred to Table 13.2 (which constitute the actual IHA) in order to be subject to are more thorough evaluation to determine whether to be regarded as Potential ignition sources and where relevant, its "effectiveness".

13.5.2 Preliminary Ignition Hazard Estimation and Evaluation

The IHA Table—has three main columns (see Tables 13.2, 13.3 and 13.4):

- **Equipment related ignition sources** (Main column 1, presented in Table 13.2) which receives its content from Table 13.1, and are further divided in three columns (1a, 1b and 1c): Source, Cause description and Basic likelihood.
- **Preventive measures applied** (Main column 2, presented in Table 13.3) which specifies the ignition source preventive measures to be implemented, and is further divided in two columns (2a and 2b): Description and Documentation.
- **Eventual assessment** (Main column 3, presented in Table 13.4) which contains the conclusion and the final categorisation of the equipment, and is further divided in three columns (3a, 3b and 3c): Likelihood, Category and Restrictions.

In this second step of the IHA process, the main Column "Equipment related ignition sources" is set into focus, here presented with a few words (see also example in Chap. 14).

For each identified equipment related ignition source (The triggering of one type of possible ignition source may be due to several causes), the underlying cause is described in sub-column 1b. Its initial estimated occurrence frequency (basic likelihood without considering any possible supplementary preventive/protective

Table 13.2 Main column 1

Equipment related ignition sources (1)		
Source (1a)	Cause description (1b)	Basic likelihood (1c)
Hot surface	Hot surface of a ball bearing	Foreseeable malfunction
Hot surface	Heating of a viscosity meter	Not relevant
Mechanical spark	Due to a grinding roots pump rotor at dry run condition	Rare malfunction

measure) with which the source may become an effective ignition source is registered in sub-column 1c as:

- Normal operation, when it may occur during normal operation.
- Foreseeable malfunction, when it may occur solely due to expected malfunctions.
- Rare malfunction, when it may occur solely due to rare malfunctions.
- Not relevant, when it is assessed never to become effective.

In order to fulfil the technical documentation obligation specified in the ATEX directive 2014/34, it is recommended to supplement the evaluation in sub-column 1c with a statement regarding the reasons for the evaluated result, if not self-explanatory.

Where the likelihood of occurrence cannot be assessed, the assumption should be made, that the source is present at all times.

The frequency estimation may be difficult, especially to distinguish between expected and rare malfunctions. Some guidance taken from the literature is reproduced in Chap. 14.

The initial assessment result produced in this second IHA step is used to decide whether supplementary measures are to be defined in the following step 3 in order to achieve the equipment category aimed for.

To be classified as **Category 3** equipment requires that potential ignition sources during normal operation must be prevented.

To be classified as **Category 2** equipment requires that potential ignition sources during normal operation and expected malfunctions must be prevented.

To be classified as **Category 1** equipment requires that potential ignition sources during normal operation, expected malfunctions and rare malfunctions must be prevented.

The analysis used as an example in Table 13.2 Main column 1, points to the fact that this equipment could be classified as a Category 3 equipment. In case the equipment will be marketed with an intended use as Category 3 equipment, it may not be necessary to implement any preventive supplementary measures.

13.5.3 Preventive Measures Implemented and Their Documentation

If the assessment in step 2 of the IHA (dealt with in Sect. 13.5.2) points to the fact that the target category cannot be achieved without supplementary appropriate preventive and/or protective measures, the selected measures are described in column 2a of the IHA Table. This shall be done for each relevant potential ignition source, as illustrated below in Table 13.3 Main column 2, which is a continuation of the Table 13.2 Main column 1.

Table 13.3 Main column 2

Preventive measures applied (2)	
Description (2a)	Documentation (2b)
Bearings calculated according to ISO 281, max temperature determined under most adverse condition (100 °C)	EN 13464-1, clause 8.2 EN 13463-5, clause 6.1
	EN 13463-1, clause 5
Shock pressure resistant casing and mounting of an autonomous protective system	EN 13463-1, clause 5 EN 12874

The term "preventive and/or protective measures" covers any measure with the purpose of explosion protection: protective measures included in the EN 13463 series, but also measures during putting into service, maintenance and repair, operation, warning notices, experimental investigations providing for evidence, etc., which will decrease the probability of ignition sources becoming effective. Protective measures may thus take the form of a design feature or may also be a part of the operation instructions.

In column 2b of the IHA Table, each selected measure should be linked to the relevant documentation: References to standards, technical rules, experimental results, etc., and/or appropriate "evidence" filed in the Technical dossier (drawings, parts lists, instructions).

13.5.4 Eventual Assessment

Taking into consideration the applied preventive measure(s) for each individual ignition hazard, the reassessed likelihood of occurrence is registered in column 3a of the IHA Table, using the same terms as listed in Sect. 13.5.2, as illustrated in Table 13.4 Main column 3, which is a continuation of Table 13.3 Main column 2 of the IHA Table.

The reassessed likelihood determines the equipment category for the different identified potential ignition sources by using the categorization rules presented in Sect. 13.5.2.

Table 13.4 Main column 3

Eventual assessment (3)		
Likelihood (3a)	Category (3b) (G)	Restrictions (3c)
Rare malfunction	2	T4
Not relevant	1	T6
Not relevant	1	
Resulting category	2	T4

The resulting equipment category is finally determined by the ignition source linked to the lowest categorized group (the category representing the least protection).

In addition to the estimated likelihood and its specification of equipment category, restrictions of the intended use are often necessary, which should be included in the equipment's marking. These restrictions, to be noted in column 3c, refer most often to the temperature class or the max surface temperature. Restrictions specifying a certain Gas group (see Chap. 5) may also be relevant, or it may be appropriate to emphasize that a single substance in whose explosive atmospheres the product may be used or is not allowed to be used. It may also be relevant to draw attention to other limitations, e.g. concerning ambient temperature, ambient pressure, etc.

In line with the determination of the resulting equipment category, the final temperature specification is for example determined by the ignition source for which the highest max surface temperature is indicated.

Note: As a piece of equipment can have different interfaces to the surroundings (see Sect. 12.12 of Chap. 12) with different relevant potential ignition sources, the abovementioned IHA may have to be repeated, as appropriate.

The IHA should, as the final point, state the **Ex-marking** to be affixed to the product. At least a part of this marking is also important for the specification of compliance in the Declaration of Conformity.

13.6 Technical Documentation

The ATEX Equipment directive 2014/34 requires a technical documentation which enables the conformity of the product with the relevant requirements to be assessed. All safety related technical details must be laid down in the technical documentation.

Completing the Technical documentation does not only deal with the IHA report, but also with the protective measures derived from it. Protective measures can be available as an attribute of the construction or a design layout, as well as a part of the instructions or a specification for the manufacturing process.

The Technical documentation shall therefore include at least:

- The ignition Hazard Assessment report.
- A description of the equipment and its intended use.
- Design and manufacturing drawings, supplemented by necessary explanations as appropriate.
- Materials and their characteristics, and certificates if necessary.
- Test reports of relevant tests.
- Notified Body certificates and communications, if necessary.
- Basic instruction, operation and maintenance requirements.

Chapter 14
ATEX—Ignition Sources

14.1 Introduction

This chapter contains a listing of the possible ignition sources.

Chapter 13 gives more information concerning the distinction between the different ignition source terms (possible, equipment related, potential and effective ignition sources).

14.2 Possible Ignition Sources

The Possible ignition source (i.e. all kinds of ignition sources) table in point 1 of the Ignition Hazard Assessment scheme, presented in Chap. 13 lists the following possible ignition sources.

14.2.1 Hot Surfaces

(See also EN 1127-1, clause 5.3.2 and 6.4.2)

Explosive atmospheres can be ignited by contact with hot surfaces when the temperature of the surface reaches the atmosphere's ignition temperature. If hot surfaces can come in contact with explosive atmospheres, a safety margin should be ensured between the max surface temperature and the atmosphere's ignition temperature—see also Sect. 5.2 of Chap. 5 and Chap. 16.

Hot surfaces resulting from malfunctions are e.g. moving parts which overheat because of inadequate lubrication. Slow moving parts having a periphery speed less than 1 m/s, do normally not need protection against frictional heating.

T. Jespen, *ATEX—Explosive Atmospheres*, Springer Series
in Reliability Engineering, DOI 10.1007/978-3-319-31367-2_14

Dust deposits have an insulating effect and may inhibit the dissipation of heat into the surrounding atmosphere. This can lead to heat build-up and hence to a further rise in temperature, leading to ignition of the dust layer (equipment suited for explosive gas/air atmospheres is therefore not necessarily applicable for dust/air atmospheres).

The max surface temperature of the equipment must be established at the max ambient temperature for which the equipment is designed. The measured max surface temperature must comply with the specifications laid down in EN 13463-1/1127-1, for example in order to ensure a proper safety margin between measured and marked max temperature.

14.2.2 Flame and Hot Gases, Including Hot Particles

(See also EN 1127-1, clause 5.3.3 and 6.4.3)

Both flames and incandescent or glowing solid particles can ignite explosive atmospheres. Even very small flames are among the most effective ignition sources.

14.2.3 Mechanical Generated Sparks

(See also EN 1127-1, clause 5.3.4 and 6.4.4, and EN 13463-1, clause 6.4)

Friction, impact and abrasion processes, such as grinding, can give rise to sparking (the tearing off small particles that become hot from the energy supplied in the process). Oxidizable particles may react further with oxygen and reach even higher temperatures.

Sparks can ignite flammable gases and vapours and certain mist/air and dust/air mixtures. In deposited dust, smouldering can be caused by the sparks, and this may become the proper explosion ignition source.

Slow moving parts having a periphery speed of less than 1 m/s do normally not need protection against mechanical sparks.

According to experience, mechanically generated sparks seem to be unable to ignite coal/dust-air mixtures during normal operation of coal grinding.

Impact and abrasion processes involving rust and light metals (e.g. aluminium and magnesium) and their alloys may initiate an aluminothermic (thermite) reaction which can give rise to particularly incendive sparking.

14.2.4 Electrical Equipment

(See also EN 1127-1, clause 5.3.5 and 6.4.5)

Even at low voltages, electrical sparking and hot surfaces may occur as a source of ignition in electrical apparatus (e.g. on making and breaking circuits, by loose connections and as a result of stray electric currents), including mobile telephones.

Electrical equipment must always be regarded as a possible ignition source and relevant standards of the EN 60079 series should be consulted.

14.2.5 Static Electricity

(See also EN 1127-1, clause 5.3.7 and 6.4.7, and EN 13463-1, clause 6.7)

Static electricity is the electrical charging of materials through physical contact and separation and the positive and negative electrical charges formed by this process. If the process is not or cannot be properly grounded, allowing the charge build-up to be safely dissipated, the charge may build up to the point where it will discharge with a static arc, which may provide an ignition source.

The importance of grounding may be illustrated by the fact that most hydrocarbon gases and vapours have minimum ignition energies in the range 0.1–1 mJ, and a person walking across a carpeted floor can develop a potential difference large enough for a 40 mJ discharge.

14.2.6 Chemical Reactions, Including Self Ignition and Spontaneous Ignition

(See also EN 1127-1, clause 5.3.14 and 6.4.14)

Chemical reactions, which develop heat (exothermic reactions) can cause substances to heat up and thus become a source of ignition (when the rate of heat generation exceeds the rate of heat loss to the surroundings).

Among the decisive parameters are the volume/surface ratio of the reacting system, the ambient temperature and the residence time. The high temperatures developed can lead to the initiation of smouldering and/or burning and also to the ignition of explosive atmospheres.

Coal is susceptible to spontaneous combustion initiated by the heat released by the exothermic reaction between carbon and oxygen. The self-ignition may start a smouldering combustion where the oxidation occurs at the surface of the solid fuel. A smouldering combustion entails a supplementary risk of CO emissions (toxic and flammable gas) because of incomplete combustion.

14.2.7 Lighting

(See also EN 1127-1, clause 5.3.8 and 6.4.8)

Direct lighting can easily ignite explosive atmospheres, and lighting can create stray currents, which may produce dangerous sparks.

14.2.8 Stray Electric Currents e.g. from Cathodic Corrosion Protection Systems

(See also EN 1127-1, clause 5.3.6 and 6.4.6)

Stray currents may give rise to electric sparks and cause heating of their paths, both of which may ignite explosive atmospheres.

14.2.9 Radio Frequency Waves

(See also EN 1127-1, clause 5.3.9 and 6.4.9)

Powerful fields e.g. from radio transmitters or generator for welding may under certain circumstances cause ignition in explosive atmospheres.

14.2.10 Optical Radiation, Including Laser Radiation

(See also EN 1127-1, clause 5.3.10 and 6.4.10)

Radiation, as sunlight, in the optical range may bring about an ignition when focused by objects e.g. glasses.

Possessing more energy, laser radiation may cause high temperature rises and act as an effective ignition source.

14.2.11 Ionising Radiation

(See also EN 1127-1, clause 5.3.11 and 6.4.11)

Equipment generating ionising radiation may ignite explosive atmospheres by self-heating, chemical reactions and absorption, e.g. by dust particles, of the emitted energy.

14.2.12 Adiabatic Compression and Shock Waves

(See also EN 1127-1, clause 5.3.13 and 6.4.13)

When gases are compressed, heat is generated. If the rate of heat generation exceeds the rate of heat loss to the surroundings, the temperature of the system will rise. When the rate of compression is very rapid, heat loss becomes negligible, resulting in "adiabatic compression". A compression ratio of 18:1 for air initially at ambient temperature produces temperatures of 674 °C.

Shock waves may be generated by the release of high-pressure gases in piping, which may result in high temperature rises.

14.3 Expected and/or Rare Malfunction

When a Category 1 classification is aimed for, sub-column 1c (see Table 13.2) and sub-column 3a (see Table 13.4) of the Ignition hazard assessment table presented in Chap. 13, Sect. 13.5, requires expert assessments. Here, you have to distinguish between an expected malfunction and a rare malfunction or between a rare malfunction and scenarios that is outside the scope of the ATEX equipment directive. Since no quantitative definitions exist, this may be a difficult task.

The following information may give some guidance:

- Fault statuses known from practice, i.e. fault statuses, which have already occurred, must certainly be classified as expected malfunctions.
- Faults which, admittedly, are basically conceivable but which have not occurred to date in practice in the case of a large number of applications should certainly be categorised as rare malfunctions. However, in this case it must be kept in mind to what extent feedback is received on the actual application conditions of the products and whether this is systematically registered, recorded and evaluated in order to provide objective evidence. When in doubt, one should opt to be on the safe side, i.e. classification as an expected malfunction.
- If an ignition source can take effect only when two expected malfunctions occur simultaneously, this must be categorised as a rare malfunction. With this in mind, all possible combinations of two expected malfunctions must be reviewed in order to establish whether they can occur simultaneously and whether there is an ignition hazard, which has not been registered to date.
- Combinations of one expected malfunction and one rare malfunction or even less probable combinations (three faults, etc.) do not need to be taken into consideration.

14.4 Examples on the Use of the Ignition Hazard Assessment Scheme

14.4.1 Possible Ignition Source Table

In the first step of the Ignition hazard assessment scheme (see Chap. 13, Sect. 13.5.1), the 12 possible ignition sources are listed in Table 13.1.

In column 2 of the table it shall be indicated if the source is present as an equipment related source or not. It may be relevant to specify the existence (or non-existence) of the source in relation to one or more interfaces between the equipment and the surroundings—see Chap. 13.

Column 3 shall contain a short description of the circumstances that may trigger the source or "not present" when the source is not relevant at all. If the source may be triggered only by foreseeable process conditions, this fact may be registered as "Foreseeable process conditions". An example on the use of Table 13.1 is presented in Table 14.1.

Table 14.1 Example on the use of Table 13.1

Possible ignition source	Equipment related (yes/no)	Comments
Hot surfaces	Yes	Inside and outside—vane friction, particle ingress
Flame and hot gases (including hot particles)	Yes (inside) No (outside)	Compression temperature to be measured
Mechanical sparks	Yes	Particles could produce hot-spots
Electrical sparks	No	Not present
Static electricity	Yes	Vanes, lip seal, air filter, float valve
Chemical reactions (including self ignition)	Yes	Possible with process fluid/gas
Lightning	No	Not present
Stray electric currents and cathodic corrosion protection	No	Not present
Radio frequency waves	No	Not present
Optical radiation (including laser radiation)	No	Not present
Ionizing radiation	No	Not present
Adiabatic compression and shock waves	No	Not present

Table 14.2 Example covering first main column

Potential ignition sources		
Source	Cause description	Basic likelihood
Hot surface	Hot surface of a ball bearing	Foreseeable malfunction
Hot surface	Heating of a viscosity meter	Not relevant
Mechanical spark	Due to a grinding roots pump rotor at dry run condition	Rare malfunction

14.4.2 IHA Table

The following examples on the use of the "Ignition hazard assessment table" in the second step of the Ignition hazard assessment scheme is split into three tables here due to lack of space on the page—equivalent to the presentation in Chap. 13.

Table 14.2 covers the first Main column with three sub-columns, Table 14.3 covers the second Main column with two sub-columns, and finally Table 14.4 covers the third Main column with three sub-columns.

Table 14.3 Example covering second main column

Preventive measures	
Description	Documentation
Bearings calculated according to ISO 281, max temperature determined under most adverse condition (100 °C)	EN 13464-1, clause 8.2 EN 13463-5, clause 6.1 EN 13463-1, clause 5
Shock pressure resistant casing and mounting of an autonomous protective system	EN 13463-1, clause 5 EN 12874

Table 14.4 Example covering third main column

Eventual assessment		
Likelihood	Category	Restrictions
Rare malfunction	2G	T4
Not relevant	1G	T6
Not relevant	1G	
Resulting category	2G	T4

Chapter 15
ATEX—Communicating IHA Results

15.1 Communicating IHA Results

Equipment possessing own effective ignition sources intended for use in potentially explosive atmospheres are regulated by ATEX Equipment directive 2014/34, which means that the equipment must be CE- and Ex-marked according to the outcome of the Ignition hazard assessment (and be accompanied by an ATEX Equipment directive 2014/34 conformity declaration)—see Chap. 16.

Where the Ignition hazard assessment has revealed potential ignition sources due to foreseen process conditions, this fact must always be communicated to the user through the accompanying instructions, informing him about the need to take additional precautions.

But when the delivery includes more than equipment intended for use in (surrounded by) potentially explosive atmospheres, it may not be sufficient just to inform about the hazards to the customer. Such situations clearly arise when the equipment becomes an integrated part in processing material-flows where potentially explosive atmospheres may be formed (like a coal mill installation).

In these cases, effective ignition sources must be eliminated or—if elimination is not possible—the process equipment must be designed and constructed to limit the effects of an explosion to a safe level. Here the Ignition hazard analysis is part of the overall risk assessment of the equipment/process, which is not directly communicated to the customer, but the results will be manifested through the instruction materials. Should the equipment have own ignition sources, the equipment falls under the ATEX Equipment directive 2014/34 and the result of the Ignition hazard assessment shall be reflected in the Ex-marking.

The next part of this section deals with the situation where no inherent effective ignition sources have been demonstrated for equipment intended for use in potentially explosive atmospheres.

T. Jespen, *ATEX—Explosive Atmospheres*, Springer Series in Reliability Engineering, DOI 10.1007/978-3-319-31367-2_15

15.2 Communicating Absence of Inherent Potential Ignition Sources

According to ATEX Workplace directive 1999/92, the employer shall classify places where an explosive atmosphere may occur into zones.

Furthermore the employer shall select and use equipment classified for use in the appropriate zones.

Classification of equipment in categories is regulated by ATEX Equipment directive 2014/34 which does not include in its scope equipment without own ignition sources (but protective systems are included regardless of possible ignition sources).

Strictly compliance with ATEX directive 2014/34 means therefore that equipment possessing no inherent (without taking into consideration any supplementary preventive or protective measures) potential ignition sources do not fall under the directive, and should not be CE- and Ex-marked according to this directive.

Where the absence of inherent potential ignition sources have been substantiated based on a sound ignition hazard assessment, such equipment—maybe with some limitations—can of course be used in areas where explosive atmospheres may occur.

This fact may be of major importance to communicate to potential customers because when they endeavour to comply with ATEX Workplace directive 1999/92, they may restrict buying equipment to what has been classified according to ATEX Equipment directive 2014/34, which in principle does not cover equipment without own effective ignition sources.

This problem may be tackled in one of three ways.

15.2.1 Using ATEX Equipment Directive 2014/34 Anyhow

Although the equipment possesses no inherent effective ignition sources, it is judged acceptable to proceed as if the equipment falls under ATEX Equipment directive 2014/34.

You may therefore carry out a sound Ignition hazard assessment, comply with the appropriate assessment procedure, CE- and Ex-mark the equipment, fill out the ATEX EU Declaration of Conformity and make sure correct instructions have been elaborated and that other elements of the Technical dossier are present.

Note: When a Category 1 or 2 classification is aimed for, this solution implies the involvement of a Notified Body (at least to receive the Technical documentation for Category 2 equipment).

15.2.2 Supplementing a Required Declaration of Conformity

In case of equipment which must comply with other directives, e.g. the Machinery directive 2006/42, the required Declaration of Conformity can be supplemented by a sentence declaring that the equipment has been examined according to ATEX equipment directive 2014/34, which has demonstrated no inherent potential ignition sources. The equipment may therefore be used in Zone X or Zones XY (to be specified according to what can be guaranteed), when operated and maintained according to the instructions.

15.2.3 Using a Separate Declaration

An independent declaration of absence of inherent potential ignition sources may be the most appropriate manner to inform customers of this fact.

The following model may be used:

Declaration on absence of inherent potential ignition sources According to ATEX directive 2014/34	
Manufacturer	XXX
Hereby declares that	
Product	(Equipment description)
Identification No.	(Appropriate identification number(s))
Has been subject to an Ignition Hazard Assessment according to EN 13463-1 demonstrating no inherent potential ignition sources, when operated and maintained according to the accompanying instructions.	
The equipment may therefore be used in the following areas where explosive atmospheres may occur: (The relevant zone(s) is indicated).	
Taking into consideration the following restrictions: (e.g. regarding temperature restrictions and/or process conditions as specified in the Ignition Hazard Assessment table of the Ignition Hazard Assessment scheme)	
Signed by: (including place and date)	

Chapter 16
ATEX—Ex-Marking

16.1 Ex-Marking

The ATEX equipment and protective systems Ex-marking shall in principle provide all the necessary information for an appropriate and correct choice of product, and for its safe use or installation in ATEX zones.

The ATEX markings that are used in practice may be rather confusing because they are not really "standardised", and the recommendations in relevant standards are not aligned with the classification scheme of the ATEX Equipment directive 2014/34. But in any case, the specific product marking must be explained in the accompanying user instructions.

This chapter gives an overview of the ATEX directive 2014/34 legal requirements to the marking (CE- and Ex-marking) of ATEX products, which are comparable with the marking specification in EN 13463-1.

Subsequently, the marking specifications presented in the fundamental ATEX standard (EN 60079-0) on electrical equipment—using the basic parameter "Equipment Protection Level" (EPL) will be presented. These marking specifications are reiterated in the draft standard prEN ISO 80076-36 on "Non-electrical (mechanical) equipment for use in explosive atmospheres—Basic methods and requirements".

16.2 CE- and Ex-Marking According to ATEX Directive 2014/34

As for machinery and many other products, equipment, protective systems and devices falling under the ATEX equipment directive 2014/34 must bear the CE-marking and be accompanied by an EU Declaration of compliance in order to demonstrate compliance with the directive requirements. This is a prerequisite for

T. Jespen, *ATEX—Explosive Atmospheres*, Springer Series
in Reliability Engineering, DOI 10.1007/978-3-319-31367-2_16

these items to be placed on the market or put into service in EU (and EEA = European Economic Area).

Components, as defined by the Directive, must not have the CE-marking affixed—unless this should be required by another EU directive covering the component in question.

On ATEX equipment the CE-marking shall be followed by the identification number of the Notified Body where such a body is involved in the Conformity assessment procedure in the production stage—see Chap. 12, Sect. 12.7.

Furthermore, equipment and protective systems under ATEX directive 2014/34 must be marked with the following information:

- Name, registered trade name/mark and address of the manufacturer
- Designation of series or type
- Batch or serial number, if any
- Year of construction
- Specific Ex-marking

Where applicable, the identification number of the Notified Body may be followed by any other mark indicating a special risk or use.

The specific Ex-marking starts with symbol.

This symbol is to be followed by the specification of equipment Group (I or II) and Category:

- For Group I: M1 or M2
- For Group II: 1, 2 or 3

Concerning Group II, this is to be followed by:

- the letter "G" when the equipment is for use in gas (gases, vapours or mists) explosive atmospheres,

and/or

- the letter "D", when the equipment is for use in dust explosive atmospheres.

Note: "GD" means suitable for both gas and dust explosive atmospheres.

Examples: Ex II 2G, Ex II 3GD, Ex I M2 and Ex (For a protective system for use in gas and dust explosive atmospheres).

In many cases, the specification of more than one category is relevant for a Group II product, because its interface to for example the process flow and its external surroundings fulfils different category requirements. In such cases, it is recommended to use the following format (X and Y indicate different Group II categories):

- "X/Y" which means the product has two interfaces fulfilling different category requirements: X and Y respectively.
- "X/-" which means that a part of the products interface fulfils category X requirements, while another part is not in compliance with the directive and therefore not intended to be surrounded by a potential explosive atmosphere.

Furthermore it is recommended that devices covered by the Directive and separately placed on the market, shall be marked with the category of the equipment under control in round brackets, and such devices which contain an own potential ignition source intended for use in a potential explosive atmosphere shall be marked as equipment.

Examples:

- Ex II (1) GD: Device outside zone classified areas which can be connected e.g. to Category 1 equipment.
- Ex II 2(1) G: Category 2 equipment containing a safety device for a Category 1 equipment.

Concerning **components**, the Directive requires these items to be accompanied by an Attestation of conformity which shall state their characteristics and inform about incorporation, but no marking requirement is specified. On the other hand, stating their characteristics generally resembles Ex-markings as e.g. Ex II 1 GD c Tx.

Above-mentioned marking requirements are the legal EU minimum requirements set by the ATEX Equipment directive 2014/34, which may be supplemented by any information which is necessary for ensuring the safe use of the product.

Relevant supplementary information is in general the max surface temperature, expressed as the numerical temperature value (used for dusts) or through the temperature code (used for gases), and the Ex-code for the selected protection type—see Chap. 5.

Especially for electrical equipment, the specification of IP (Ingress Protection) may be relevant—see Chap. 5, Sect. 5.4.

16.3 Ex-Marking According to EN 13463-1

EN 13463-1 that is the basic ATEX standard for non-electrical (mechanical) equipment contains supplementary specifications regarding the Ex-marking of equipment:

- Group II equipment having a max surface temperature greater than 450 °C shall bear only the inscription of the temperature (No temperature code).
- Group II equipment designed and marked for use in a particular gas, does not need a temperature marking.
- Where the actual max surface temperature does not depend on the equipment itself, but mainly on operating conditions, the temperature marking is done by "TX", for which the instructions for use shall give the relevant information.
- If special conditions for the safe used applies, the symbol "X" shall be used in the marking and explained in the instructions, unless the use of a warning marking can be used to provide appropriate information.

Marking examples:

- II 1G IIB T4 c/k: Group II, Category 1 equipment, suitable for gas belonging to Gas Group IIB, having max surface temperature of 135 °C, utilizing protection by construction and liquid immersion applied to the same ignition source.
- II 2G IIB c k T4: Group II, Category 2 equipment, suitable for gas belonging to Gas Group IIB, having max surface temperature of 135 °C, utilizing protection by construction and liquid immersion applied to different ignition sources.
- II 2D c T110°C: Group II, Category 2 equipment, suitable for dust, utilizing protection by construction, having a max surface temperature of 110 °C.
- II 2G d T3/1G c T2 X: Group II equipment having two Categories (e.g. for different parts of the equipment): A Category 2 suitable for gas, having a max surface temperature of 200 °C, utilising protection by flameproof enclosure, and a Category 1 suitable for gas, having a max surface temperature of 260 °C, utilizing protection by construction.

16.4 Ex-Marking According to EN 60079-0

EN 60079-0 that is the basic ATEX standard for electrical equipment contains marking specifications that categories the equipment in a slightly different way compared to ATEC Equipment directive 2014/34, using the parameter "Equipment Protective Level" (EPL).

The draft (parallel) standard prEN ISO 80079-36 for non-electrical equipment—foreseen to supersede EN 13463-1—contains the same approach to marking of ATEX equipment.

Chapter 5, Sect. 5.6 contains a comparison between the marking specification of ATEX directive 2014/34 and EN 60079-0, see Table 16.1.

The most noteworthy difference seems to be as where the Directive operates with two equipment Groups (I and II), the standard splits the directive Group II into a Group II with equipment suitable for gas, vapour and mists, and a Group III with equipment suitable for dusts.

Table 16.1 Marking according to EN 60079-0 and directive 2014/34

EN 60079-0		ATEX directive 2014/34		
EPL	Equipment group	Equipment group	Category	Zone
Ma	I	I	M 1	NA
Mb			M 2	
Ga	II	II	1 G	0
Gb			2 G	1
Gc			3 G	2
Da	III		1 D	20
Db			2 D	21
Dc			3 D	22

The EPL levels are just another way of expressing the combination of potential explosive atmosphere type (gas, vapour, must on the one side and dust on the other) and level of prevention of potential ignition sources (For Group II identified by Category 1, 2 or 3).

With the intention of fulfilling both the ATEX legislation and the standard specifications, the subsequent Sect. 16.5 describes a marking that includes "all"—which means that the marking becomes somewhat complicated to understand, but you will meet it on some ATEX electrical equipment in a more or less equivalent form.

16.5 A Comprehensive ATEX Equipment Marking

With a view to be in compliance with the obligatory marking requirements of ATEX directive 2014/34—taking the relevant international standard specifications into consideration—the following marking system has been suggested. It consists of two parts: A first part fulfilling ATEX legislative requirements, and a second part fulfilling standard specifications (which can be said, is also partly required by the legislation):

An example

CE_{XXXX} Ɛx II 2 GD EEx d IIB T4 IP66 Gb Db

Where:

- CE_{XXXX} indicates the obligatory CE-marking accompanied by the identification number of the Notified Body involved in the production control phase **when** this involvement is required by the directive (voluntary certification should not lead to the affixing of the number of the Notified Body, in order to avoid confusion).
- Ɛx (which should be in a hexagon) introduces the obligatory ATEX Ex-marking consisting of Group (I or II), Category (1, 2 or 3) and type of explosive atmosphere: G (gas), D (dust) or GD (both gas and dust).

- EEX introduces the standard specification for explosion protection marking consisting of the indication of protection type (the Ex code, here "d"), gas group (here IIB), max surface temperature (here T4 = 135 °C), IP code and EPL level(s).

What is said above concerning:

- The use of "/" to specify that the interfaces of the equipment belongs to more than one Category, as e.g. 2/–,
- The use of "Tx" when the max surface temperature is mainly dependent on non-equipment external factors, and
- The use of "X" when there special circumstances of key importance for the safe use of the equipment,

are also of relevance regarding the abovementioned combined equipment marking, **although** the eventual equipment Ex-marking may end up being something that only special ATEX experts can understand (Anyhow, the marking must be explained in the instruction for use).

Part III
Basic Fire and Explosion Aspects

Chapter 17
Fires and Explosions—Basic Knowledge

17.1 Introduction

This chapter contains some "basic" information especially on explosions in order to provide a basis for performing some preliminary risk assessments and for understanding discussions and other texts on fire and explosion risks.

First of all it should be emphasized that "explosions" are an extremely difficult subject and still characterized by a lack of scientific knowledge. Therefore when one moves outside what could be called "the fundamental knowledge" one may encounter experts with somewhat divergent views.

17.2 Definitions

Note: Similar and slightly different definitions may be found in the literature.

Combustion: Burning of gas, liquid or solid (fuels) in which the fuel is oxidised resulting in heat release (increased temperature) and often light emission.

Combustible dust: Combustible particulate solid that presents a fire or explosion hazard when suspended in air or some other oxidizing medium over a range of concentrations, regardless of particle size or shape.

Fire: Combustion where fuel and oxygen is mixed during the combustion process (opposite to premixed fuel and air which may allow the fuel to burn much faster e.g. as an explosion).

Flash fire: A flash fire is the term for a slow deflagration of a premixed truly unconfined, unobstructed fuel cloud producing negligible overpressure.

Flash point: The minimum temperature at which a liquid gives off sufficient vapour to form a flammable mixture with air, near the surface of the liquid or within the vessel used.

T. Jespen, *ATEX—Explosive Atmospheres*, Springer Series
in Reliability Engineering, DOI 10.1007/978-3-319-31367-2_17

Fire triangle: The three essential components (which have to be present simultaneously) to create a fire: Fuel, oxygen (air) and effective ignition source.

Deflagration: Rapid burning with combustion wave propagating at subsonic velocity (speed lower than the speed of sound), driven by the transfer of heat.

Detonation: Supersonic (wave having a speed higher than the speed of sound) combustion in which a flame and shock wave is propagating forward due to energy release with an extremely intense variation of pressure.

Explosion: Violent chemical reaction where the temperature and pressure are raised very fast. It may be as a deflagration or a detonation with devastating pressure increases (primarily because of temperature increases and not due to a change in moles).

- As an example, 500 g/m^3 coal dust uniformly dispersed in a 20 l container at 300 K, combust completely in 0.14 s with a temperature raise to about 2100 K and a pressure raise to about 7 times the starting pressure.

Gas explosion: An explosion—a deflagration or a detonation—of a premixed combustible gas cloud.

Dust explosion: An explosion—a deflagration or a detonation—of a premixed (dispersed) combustible dust cloud (in a confined space).

Explosion triangle: The three essential components (which have to be present simultaneously) to create an explosion: Fuel, oxygen (air) and effective ignition source.

To represent the gas explosion situation one has to add the following 3 conditions:

- High degree of dispersion (mixing)
- Concentrations within explosion limits
- Hazardous quantities.

Explosion pentagon: The five essential components/conditions (which have to be present simultaneously) to create a dust explosion: Fuel (combustible dust), oxygen (air), effective ignition source, dispersion (premixed) and confinement.

To represent the dust explosion situation one has to add the following 2 conditions:

- Concentrations within explosion limits
- Hazardous quantities.

Outside confinements, dust combustion in a premixed cloud may be characterized as a deflagration—but without developing harmful overpressures.

Smouldering: Slow, low-temperature, flameless combustion occurring on the surface of a solid, sustained by the heat evolved. The solid may be coal, cellulose, wood (sawdust), cotton or synthetic foams.

Pyrolysis: Chemical decomposition of a condensed substance by heating.

Confined explosion: Explosions within tanks, process equipment, pipes, closed rooms and in underground installations.

Partly confined explosions: Explosions occurring when a fuel is accidentally released inside a building which is partly open.

Unconfined explosions: Gas explosions in open areas (A truly unconfined, unobstructed gas cloud ignited by a weak ignition source will only produce small overpressure while burning).

Turbulence: In fluid dynamics, the flow is divided into laminar and turbulent regimes. The turbulence is very important for how fast the flame can propagate in a premixed fuel cloud. The turbulence will wrinkle the flame front and increase diffusion of heat and mass and thereby cause higher burning rate.

17.3 Fires and Explosions

When combustible gas or evaporating liquid is emitted into the atmosphere and the gas cloud formed from the release, is not within flammability limits or if an ignition source is lacking, the gas cloud may be diluted and disappear.

In case of immediate ignition (i.e. before mixing with air) a fire will occur.

If a large combustible premixed fuel-air cloud is formed and ignited, an explosion may occur. The pressure generated by the combustion wave depends on how fast the flame propagates and how the pressure can expand away from the fuel cloud (determined by degree of confinement and pressure relief).

The flame can propagate through the cloud in two different modes:

- By deflagration, or
- By detonation.

The **deflagration** mode is most common. Here the flame propagates at a speed less than the speed of sound (subsonic) relative to the unburned gas immediately ahead of the flame (Relative to a stationary observer, typically from 1 up to 1000 m/s). In non-confined areas, the pressure may reach values of 4–5 times the starting pressure.

The **detonation** mode is characterized by a flame propagating at a speed higher than the speed of sound (supersonic). The shock wave and the combustion wave are coupled and propagate at velocities of 1500–2000 m/s, and the peak pressure is typically 15–20 bar.

In an accidental gas explosion of a hydrocarbon-air cloud ignited by a weak source, the flame will normally start out as a slow laminar flame with a velocity of the order of 3–4 m/s. If the cloud is truly unconfined and unobstructed, the flame is not likely to accelerate to velocities of more than 20–25 m/s, and the overpressure will be negligible if the cloud is not confined.

In a building with some process equipment, the flame may accelerate to several hundred meters per second. The unburned gas is therefore pushed ahead of the flame and a turbulent flow is generated. When the flame propagates into a turbulent field, the effective burning rate will increase and the flow velocity and turbulence ahead of the flame will increase further. This strong positive feedback mechanism is causing flame acceleration and high explosion pressures and in some cases transition to detonation.

In a confined situation, such as a closed vessel, a high flame velocity is not a requirement for generation of pressure. In a closed vessel there is no or very little relief (i.e. venting) of the explosion pressure and therefore even a slow combustion process will generate pressure.

The consequences of an explosion will depend on:

- Type of fuel and oxidiser
- Size and fuel concentration of the combustible cloud
- Location of ignition point
- Strength of ignition source
- Size, location and type of explosion vent areas
- Location and size of structural elements and equipment
- Mitigation schemes
- For dust explosions: Particle size distribution and moisture.

Explosion may be very sensitive to changes in these factors. Therefore it is not a simple task to estimate the consequences of an explosion.

17.4 Deflagrations

A deflagration is the most common mode of flame propagation in accidental explosions. The flame speed ranges from a few m/s up to 500–1000 m/s. The pressure ranges from a few mbar to several bars, depending on the flame speed.

When a gas cloud is ignited by a week ignition source (i.e. spark or hot surface) the flame starts as a laminar flame. For a laminar flame the basic mechanism of propagation is molecular diffusion of heat and mass. This diffusion process of heat and mass into the unburned gas is relatively slow and the laminar flame will propagate with a velocity of the order of 3–4 m/s.

In most accidental explosions the laminar flame will accelerate and transit into a turbulent deflagration (i.e. a turbulent flame), since the flow field ahead of the flame front becomes turbulent. The turbulence is caused by the interaction of the flow field with process equipment, piping, structures, etc. One of the mechanisms causing the increased burning rate in turbulent deflagrations is the wrinkling of the flame front by large turbulent eddies.

A flame velocity of the order of 100 m/s is required to produce pressure waves of a significant strength (around 0.1 barg) in unconfined clouds.

If the explosion happens inside a closed vessel, fast flame propagation is not required to obtain high pressures. A stoichiometric fuel-air cloud in a closed vessel will give up to 8–9 bar when exploding.

In a partly confined area with obstacles (i.e. process equipment, piping, etc.) the flame may accelerate to several hundred meters per second during a gas explosion. The mechanisms causing the increased burning rate in turbulent deflagrations are the wrinkling of the flame front by large eddies and the turbulent transport of heat and mass at the reaction front. This turbulence is mainly caused by the interaction of the flow with structures, pipe racks, etc.

The flame acceleration can to some extent be avoided by venting the hot combustion products. The flow and turbulence in the unburned mixture ahead of the flame will be reduced. Venting combustion products is a very effective way of minimising the acceleration effect of repeated obstacles.

Venting of unburned gas ahead of the flame may also contribute to a lower explosion pressure, in particular when the venting directs the flow away from repeated obstacles. If venting unburned gas leads it past repeated obstacles, flame acceleration will most likely occur.

When a deflagration propagates through a region of obstacles and then ends up in an unobstructed region, the flame speed will normally drop and adjust to the new environment.

The consequences and the probability of occurrence of gas explosions will to a large extent depend on fuel type. Under similar experimental conditions different fuel-air mixtures will generate different explosion pressures. At the present time there are no standard procedures for classifying the explosion hazard regarding pressure generation for different fuels. However, various experiments with turbulent deflagrations and detonations show that common fuels can be ranked, at least qualitatively.

Hydrogen gave the highest explosion pressure: 8 barg. Hydrogen and acetylene are the two most reactive fuels that are normally handled. Ethylene is also very reactive: 2 barg. Propane and ethane are somewhat less reactive and seem to form an intermediate level of explosivity: 0.5 barg. Methane is the least reactive fuel: 0.2 barg. Many other hydrocarbons (e.g. butane) fall into the same group as propane and ethane.

Other substances that are not normally regarded as fuels, like ammonia (NH_3) can also cause explosions. Ammonia burns very slowly, but in a confined situation, it can cause serious explosions.

For more heavy hydrocarbons such a heptane, the fuel may be dispersed into the air as a mist. Experiments on mist-air mixtures are limited, but experiments have revealed that the reactivity of alkane mist-air mixtures is in the same order of magnitude as propane-air mixtures.

The reactivity of natural gas is comparable to that of equivalent methane-propane mixtures.

For single fuels the maximum explosion pressure is normally observed at stoichiometric or slightly rich mixtures.

The concentration range causing significant explosion pressures is depending on the geometry where the explosion is occurring. The more confined and obstructed the geometry, the wider the concentration range. The limits for explosion pressure must not be confused with explosion limits (or flammability limits).

Both the strength and the location of the ignition source can be important factors in determining the consequences of a gas explosion.

Even though extreme ignition scenarios exist, the most likely ignition source is a weak ignition like a hot surface or a spark. In consequence analysis it is common to choose a weak ignition source as a probable scenario.

Various experiments and simulations have shown that explosion pressures can be very sensitive to the location of the ignition point. In many scenarios the peak explosion pressure can be changed by an order of magnitude if the ignition location is moved from “worst case” to a more favourable place. In general the lowest pressure is obtained if the ignition point is:

- Close to the vent area, or
- At the edge of the cloud

but exceptions exist.

17.5 Confined (Internal) Explosions

For internal explosions the fuel cloud size is the main parameter determining pressure build-up. The geometrical conditions nearly always support flame acceleration and pressure build-up.

Normally one can assume that the flame is a slow deflagration with a velocity of less than 20 % of the initial speed of sound, hence local high pressures due to high burning rate is neglected. For a slow deflagration in a homogeneous gas mixture, the pressure in the vessel will gradually increase as the flame consumes the gas mixture, For most hydrocarbon fuels, a stoichiometric fuel-air cloud with initial pressure of 1 atm will give 8–10 bar pressure when burning under constant volume conditions. The highest pressure is found for slightly rich mixtures. When the fuel concentration approaches the flammability limits, the explosion pressure will be reduced, but even close to the flammability limits, the theoretical values for constant volume combustion will be in the 4–5 range. Even a cloud near the flammability limit can, in an explosion, cause significant pressure build-up in a closed vessel.

Some experts state that detonations occur in pipelines, but are very improbable in vessels. In an empty vessel there are no obstructions causing turbulence and flame acceleration. Transition to detonation is therefore not likely in vessels, unless the gas is very detonable, the gas is large, the cloud is jet ignited or the vessel contains obstacles.

In addition to closed vessels, **pipes** (including ducts, channels and tunnels) are also typical simple geometries where internal explosion can occur. In pipes, the pressure generated by the flame has the possibility to propagate away from the combustion front. For long pipes or open ended pipes, a high flame speed is required to generate high explosion pressure. The main mechanism causing the flame to accelerate in pipes, is turbulence. When the gas burns, it expands and pushes unburnt gas ahead of the flame front. The flow ahead of the flame will cause turbulent boundary-layer to grow and the turbulence will enhance the burning rate.

In a pipe the flame can continue to accelerate until it becomes a detonation. This phenomenon is only understood in qualitative terms and cannot be predicted based on the present knowledge. The transition phenomenon is characterised by very high local pressures (pressures of 50 times the initial pressure have been measured when the transition occurred).

The run-up distance, i.e. the distance from ignition to transition to detonation in pipes is an experimental value giving some indication of the likelihood of transition to detonation. The run-up distance increases with increasing pipe diameter, and for a stoichiometric ethylene-air cloud in a 50 mm pipe it is about 4 m. On the other hand, in pipes with very small diameters the reaction may be quenched.

In a process unit one often finds that large volumes are interconnected by pipes and channels. In case of an internal explosion, these inter-connections can cause very strong pressure build-up. This phenomenon is often referred to as precompression or **pressure piling**. Pressure piling is a local dynamic effect which can cause high local explosion pressures:

- If two volumes I and II are interconnected by a pipe and a fuel cloud is ignited in volume I, the pressure will gradually increase and some unburnt gas from volume I will flow into volume II. When the flame enters volume II, the gas is precompressed and the flame has become a jet flame shouting into volume II, The pressure build-up in volume II will therefore become much higher than in volume I.

17.6 Partly Confined Gas Explosions

This section is about **gas explosions** in partly confined areas as compartments and buildings. **Dust explosions** may take place in such areas but only as slow deflagrations with negligible overpressure.

As mentions above, confinement and obstructing objects are key factors for the development of high explosion pressures in accidents. In building and partly confined areas containing process equipment, there will be confinement and obstacles. Walls, roof, floors and decks will confine the gas cloud. The process equipment and piping engulfed by the cloud will act as obstructing objects during an explosion.

The pressure build-up during a gas explosion is governed by the balance between pressure generation by the flame and relief of the pressure through venting. In an unconfined situation with large explosion vent areas, a flame speed in excess of 100 m/s is required in order to obtain damaging pressure waves. However, if a fuel-air cloud explodes within a compartment with no or very little venting, even slow burning can cause pressure build-up. In extreme cases, a slow flame can in a closed compartment cause pressures up to 8 bar—if the compartment does not disintegrate.

In an accidental gas explosion the flame will normally start out as a slow laminar flame with a velocity of the order of a few m/s. if the cloud is truly unconfined and unobstructed, the flame is not likely to accelerate to velocities of more than 20–25 m/s, and the overpressure will be negligible. In partly confined areas with some obstacles, the flame may accelerate to several hundred meters per second. The main mechanism of flame acceleration under such conditions is turbulent mixing due to generation of turbulent flow fields ahead of the flame.

If hot combustion products are vented out of a compartment, the flow and the turbulence can be reduced since the driving pressure is relieved and less gas is pushed ahead of the flame. However, in some cases one finds that pressure continues to rise even if the combustion products are vented. This is typical for compartments with small vent areas, high density of process equipment, piping, etc., or when the gas explosion reach high pressures before the combustion products starts to vent. For very high-speed deflagrations venting may not be effective at all.

A gas explosion in a partly confined space is a very complex process strongly depending on several parameters. This text therefore limits itself to a listing of these parameters:

- Shape of compartment
- Types of vent areas
- Effects of congestion and obstructions
- Strength and location of ignition source
- Gas cloud characteristics
- Deflagration to detonation possibilities
- Generation of jet flames shooting to other areas
- Mechanical ventilation system
- After explosion fires (serious fires often follows explosions)
- Mitigating effects from water deluge for fire fighting.

17.7 Gas Explosions in Process Areas and Unconfined Areas

Examples:

- 60 tons of cyclohexane was released inside the process plant in Flixborough in 1974. A large dense flammable cloud was formed and when the cloud exploded, the plant was totally demolished.
- In Port-Hudson propane was released from an underground pipeline. The release filled up a valley with propane-air mixture. The explosion started as an internal explosion in a pump-house and this triggered the unconfined cloud to detonate.

If the explosion is a deflagration, high explosion pressures will only be generated when the gas cloud is inside confined or partly confined areas or engulfs obstructing objects, such as pipework and process equipment. In a process plant the areas where high pressures can be generated by a deflagration, are mainly inside buildings, pipe bridges, in open process area where pipework and process equipment is densely packed and in tunnels and culverts. However, if the cloud detonates as a result of strong flame acceleration, the detonation will be able to propagate through the cloud without any confinement or obstructing objects.

When a fuel is accidentally released, the density of the fuel is an important parameter for the formation of the combustible cloud. When the gas is lighter than air, like hydrogen, buoyancy will make the cloud rise. In an open situation, the gas will rise and be dispersed relatively quickly. A dense gas will drift along the ground, and will not disperse as fast as a light gas. The dense gas may drift into buildings, tunnels, culverts or other confined areas. A release of a dense gas therefore has a higher potential of forming larger fuel-air clouds than a release of a light gas.

Hydrogen is a fuel that is lighter than air, disperses relative fast and causes high explosion pressures, but the size of hydrogen explosions are often fairly limited. Local damages can be severe because hydrogen is very reactive and a deflagration may accelerate very fast and easily transit into a detonation.

17.8 Flash Fires

A deflagration producing negligible overpressure is often called a "flash fire" which is characteristic for a truly unconfined, unobstructed gas cloud ignited by a weak ignition source. Under such circumstances there are no mechanisms that can accelerate the flame (i.e. a deflagration) to more than a few tens of meters per second. The combustion is so slow that burned gas will expand before any significant pressure can build up. The thermal effects are the main hazard from a truly unconfined deflagrating cloud.

However, if the same free cloud detonates due to transition to detonation in a confined neighbouring area, the result will be a very strong blast wave (air wave set in motion by an explosion).

When a fuel-air cloud burns, the hot combustion products will rise due to buoyancy (temperature caused differences in density). For a large cloud this buoyancy can be very strong and the flow ahead of the flame can even be reversed (fuel-air mixture is sucked into the fire).

A Few Facts About the Author

Educations:

- Master of Science in Chemical Engineering from the Technical University of Denmark (DTU).
- Bachelor degree in International Trade from Copenhagen Business School (CBS).

Occupations:

- 5 years as an Assistant Professor at the Institute of Chemical Industry at DTU—teaching chemical process and reaction engineering.
- 25 years filling in different positions at the Confederation of Danish Employers—providing assistance to Member Organizations and Companies concerning occupational health and safety matters, especially regarding aspects related to EU legislation.
- 9 years as Technical Adviser in a worldwide Danish engineering (and industry plant equipment supplier) company—especially regarding interlinked legal and technical matters associated with EU Product and Workplace legislation.

Other activities:

Among his other activities it can be mentioned:

- During 20 years acting as an advisory expert to the European Commission regarding Product and Workplace health and safety legislation, including the chairmanship of several advisory groups.
- During 10 years coordinated the European Employers' participation in the elaboration of EU occupational health and safety legislation concerning chemicals.
- For some years the vice-chair and employer spokesperson at the European Agency on Occupational Safety and Health (European OSHA) in Bilbao, Spain.

T. Jespen, *ATEX—Explosive Atmospheres*, Springer Series in Reliability Engineering, DOI 10.1007/978-3-319-31367-2

- Participation in several ILO activities, such as the elaboration of Conventions, Recommendations and Guidelines, for example on "Occupational safety and health management systems".

And now:
He began his retirement in the spring of 2014.

Index

T. Jespen, *ATEX—Explosive Atmospheres*, Springer Series in Reliability Engineering, DOI 10.1007/978-3-319-31367-2

The manufacturer's authorised representative in the EU is Springer Nature Customer Service Centre GmbH, Europaplatz 3, 69115 Heidelberg, Germany. If you have any concerns regarding our products, please contact ProductSafety@springernature.com

Printed and bound by CPI Group (UK) Ltd, Croydon, CR0 4YY
22/09/2026
02221071-0001